Understanding
Environmental Issues

The Open University Course Team

Prof. Andrew Blowers, OBE, Faculty of Social Sciences
(Course Co-chair and Chair of Block 4)
Dr Steve Hinchliffe, Faculty of Social Sciences (Course
Co-chair and Chair of Block 1)
Dr Dick Morris, Faculty of Technology (Chair of Block 2)
Dr Nick Bingham, Faculty of Social Sciences (Chair of
Block 3)
Dr Rod Barratt, Faculty of Technology
Dr Chris Belshaw, Faculty of Arts
Roger Blackmore, Faculty of Technology
Dr Mark Brandon, Faculty of Science
Prof. David Elliott, Faculty of Technology
Dr Joanna Freeland, Faculty of Science
Dr Wendy Maples, Faculty of Social Sciences
Dr David Morse, Faculty of Mathematics and Computing
Dr Stephen Peake, Faculty of Technology
Alan Reddish, Consultant Course Team Member
Varrie Scott, Course Manager
Dr Joe Smith, Faculty of Social Sciences
Dr Sandy Smith, Faculty of Science
Dr Charles Turner, Faculty of Science

Other Open University staff

Melanie Bayley, Editor
Pam Berry, Key Compositor
Sophia Braybrooke, Software Designer
Karen Bridge, Software Designer

Sarah Carr, Producer, BBC
Lene Connolly, Print Buying Controller
Michael Deal, QA Tester
Nigel Draper, Media Account Manager
Alison Edwards, Editor
Liz Freeman, Copublishing Adviser
Sarah Gamman, Rights Administrator
Phil Gauron, Producer, BBC
Carl Gibbard, Graphic Designer
Richard Golden, Production & Presentation Administrator
Kate Goodson, Producer, BBC
Celia Hart, Picture Researcher
Susie Hooley, Secretary, Faculty of Social Sciences
Robert Hughes, Software Designer
Lisa MacHale, Producer, BBC
Margaret McManus, Rights Assistant
Michele Marsh, Course Secretary, Faculty of Social Sciences
Lynda Oddy, QA Testing Manager
David Schulman, Producer, BBC
Lynne Slocombe, Editor
Jan Smith, Secretary, Faculty of Social Sciences
Neeru Thakrar, Secretary, Faculty of Social Sciences
Colin Thomas, Lead Software Designer
Howie Twiner, Graphic Artist
Jenny Walker, Producer, BBC
David Wardell, Media Account Manager
Amanda Willett, Series Producer, BBC
Darren Wycherley, Producer, BBC

Consultants

Dr Paul Anand, Lecturer in Economics, The Open
University
Prof. Jacquie Burgess, Professor of Geography, University
College London
Dr Noel Castree, Reader in Human Geography, University
of Manchester
Prof. Michael Drake, Professor Emeritus, Faculty of Social
Sciences, The Open University

Pam Furniss, Lecturer in Systems, Faculty of Technology,
The Open University
Prof. Pieter LeRoy, Professor in Political Sciences of the
Environment, Nijmegen University
Dr Annie Taylor, Visiting Fellow, Department of Politics,
University of Southampton
Dr Karin Verhagen, Junior Lecturer in Political Sciences of the
Environment, Nijmegen University

External assessor

Prof. Kerry Turner, CBE, Director of The Centre for Social and Economic Research on the Global Environment, University of East
Anglia

Tutor panel

Claire Appleby, Associate Lecturer, The Open University, Cambridge
Dr Ian Coates, Associate Lecturer, The Open University, Bristol
Dr Arwyn Harris, Associate Lecturer, The Open University, Wales

With thanks to
Caitlin Harvey, Course Manager, Faculty of Social Sciences
Dr Dan Weinbren, Course Manager, Faculty of Social Sciences
John Watson, Technician, Earth Sciences

Understanding Environmental Issues

edited by Steve Hinchliffe,
Andrew Blowers and Joanna Freeland

wiley.com

in association with

The Open
University

This publication forms part of an Open University course U216 *Environment: Change, Contest and Response*. The complete list of texts that make up this course can be found on the back cover. Details of this and other Open University courses can be obtained from the Course Information and Advice Centre, PO Box 724, The Open University, Milton Keynes MK7 6ZS, United Kingdom: tel. +44 (0) 1908 653231, e-mail general-enquiries@open.ac.uk

Alternatively, you may visit the Open University website at www.open.ac.uk where you can learn more about the wide range of courses and packs offered at all levels by The Open University.

To purchase a selection of Open University course materials visit the webshop at www.ouw.co.uk, or contact Open University Worldwide, Michael Young Building, Walton Hall, Milton Keynes MK7 6AA, United Kingdom for a brochure. tel. +44 (0) 1908 858785; fax +44 (0) 1908 858787; e-mail ouwenq@open.ac.uk

First published 2003 by John Wiley & Sons Ltd in association with The Open University

The Open University
Walton Hall
Milton Keynes
MK7 6AA
United Kingdom
www.open.ac.uk

John Wiley & Sons Ltd
The Atrium
Southern Gate
Chichester
PO19 8SQ

www.wileyeurope.com or www.wiley.com

Email (for orders and customer service enquiries): cs-books@wiley.co.uk

Other Wiley editorial offices: John Wiley & Sons Inc., 111 River Street, Hoboken, NJ 07030, USA; Jossey-Bass, 989 Market Street, San Francisco, CA 94103–1741, USA; Wiley-VCH Verlag GmbH, Boschstr. 12, D–69469 Weinheim, Germany; John Wiley & Sons Australia Ltd, 33 Park Road, Milton, Queensland 4064, Australia; John Wiley & Sons (Asia) Pte Ltd, 2 Clementi Loop #02–01, Jin Xing Distripark, Singapore 129809; John Wiley & Sons Canada Ltd, 22 Worcester Road, Etobicoke, Ontario, Canada M9W 1L1.

Library of Congress Cataloging-in-Publication Data

A catalogue record for this book is available from the Library of Congress.

British Library Cataloguing in Publication Data

A catalogue record for this book is available from the British Library.

ISBN 0 470 84998 3

Edited, designed and typeset by The Open University.

Printed by The Bath Press, Glasgow.

1.1

Contents

Series preface

Environment: Change, Contest and Response is a series of four books designed to introduce readers to many of the principal approaches and topics in contemporary environmental debate and study. The books form the central part of an Open University course, which shares its name with the series title. Each book is free-standing and can be used in a wide range of environmental science and geography courses and environmental studies in universities and colleges.

This series sets out ways of exploring environments; it provides the knowledge and skills that enable us to understand the variety and complexity of environmental issues and processes. The ideas and concepts presented in the series help to equip the reader to participate in debates and actions that are crucial to the well-being – indeed the survival – of environments.

The series takes as its common starting point the following. First, environments are socially and physically dynamic and are subject to competing definitions and interpretations. Second, environments change in ways that affect people, places, non-humans and habitats, but in ways that are likely to reflect differing degrees of vulnerability. Third, the unsettled, uncertain and uneven nature of environmental change poses significant challenges for political and scientific institutions.

The series is structured around the core themes of **changing** environments, **contested** environments and environmental **responses**. The first book in the series sets out these themes, and (as their titles indicate) each of the three remaining books takes one of the themes as its main concern.

In developing the series we have observed the rapidly changing world of environmental studies and policies. The series covers a range of topics from biodiversity to climate change, from wind farms to genetically modified organisms, from the critical role of mass media to the measurement of ecological footprints, and from plate tectonics to global markets. Each issue requires insights from a variety of disciplines in the natural sciences, social sciences, arts, technology and mathematics. The books contain chapters by a range of authors from different disciplines who share a commitment to finding common themes and approaches with which to enhance environmental learning. The chapters have been read and commented on by the multidisciplinary team. The result is a series of books that are unique in their degree of interdisciplinarity and complementarity.

It has not been possible to include all the topics that might find a place in a comprehensive coverage of environmental issues. We have chosen instead to use particular examples in order to develop a set of themes, concepts and questions that can be applied in a variety of contexts. It is our intention that as you read these books, individually or as a series, you will find thought-provoking and innovative approaches that will help you to make sense of the issues we cover, and many more besides. From the outset our aim has been to provide you with the equipment necessary for you to become a sceptical observer of, and participant in, environmental issues.

In the series we shall talk of both 'environment' and its plural, 'environments'. But, what do these terms mean? As a working definition we take 'environment' to indicate surroundings, including physical forms and features (land, water, sky) and living species and habitats.

'Environments' signifies spaces (areas, places, ecosystems) that vary in scale and are connected to, are a part of, a global environment. For instance, we may think in terms of different environments, such as the Scottish Highlands, the Sahara Desert or the Siberian taiga (forest). We may think in terms of components of environments, for instance housing, industry or infrastructures of the built environment. The phrase 'natural environments' may be used to emphasize the contribution of non-human organisms and biological, chemical and physical processes to the world's environment.

It is useful to keep in mind four defining characteristics of environment: environment implies surroundings; nature and human society are not separate but interactive; environment relates to both places and processes; and environments are constantly changing. These characteristics are merely a starting point; as we proceed we shall gradually reveal the relationships and interactions that shape environments.

There are several features of the books that are worthy of comment. While each book is self-contained, you will find references to earlier and later material in the series depicted in bold type. Some of the cross-references will also be highlighted in the margins of the text so that you may easily see their relevance to the topic on which you are currently engaged. The margins are also used in places to emphasize terms that are defined for the first time.

Another feature of the books is the interactivity of the writing. You will find questions and activities throughout the chapters. These are included to help you to think about the materials you are studying, to check your learning and understanding, and in some cases to apply what you have learned more widely to issues that arise outside the text. A final feature of the chapters is the summaries that appear at the end of each major section, to help you check that you have understood the main issues that are being discussed.

We wish to thank all the colleagues who have made this series and the Open University course possible. The complete list of names of those responsible for the course appear on an earlier page. Particular thanks go to: our external assessor, Professor Kerry Turner; our editors, Melanie Bayley, Alison Edwards and Lynne Slocombe; our tutor panel, Claire Appleby, Ian Coates and Arwyn Harris; and to the secretaries in the Geography Discipline, Michelle Marsh, Jan Smith, Neeru Thakrar and Susie Hooley, who have all helped in the preparation of the course. Last but not least, our thanks to our course manager, Varrie Scott, whose efficiency and unfailing good humour ensured that the whole project was brought together so successfully.

Andrew Blowers and Steve Hinchliffe
Co-chairs of The Open University Course Team

Introduction

Andrew Blowers

Environments are an inescapable part of our lives, providing the resources to sustain life. Their beauty and variety are essential to well-being, yet they also harbour destructive forces that threaten survival. We are aware of our role in conserving and in degrading them. Degradation is manifest everywhere, as environments are transformed by farming, transport, built development, tourism, industry, mining, waste disposal and dereliction. There are also the less visible, but potentially overwhelming, threats present in global climate change, in ozone layer depletion and in the development of weapons of mass destruction.

These are not matters that can be left to others to deal with, to politicians, scientific experts or environmental professionals. They are issues in which we are all, inevitably, involved. But, if we are to act responsibly, we need to know the consequences of our actions. As its title suggests, the aim of *Understanding Environmental Issues* is to help the reader develop the knowledge, skills and methods that can be applied to the study of environmental issues. This will enable us to answer the first of the book's three key questions, the question posed at the beginning of the first chapter: how can we approach the study of environments?

Chapter One is an introduction to the book and to the series. It focuses on an estuary, the Blackwater estuary in Essex, England, as a context for exploring the notion of environment. The Blackwater is an example of one environment but it is also part of other environments. We use it as a setting to show how environments and environmental issues may be understood in terms of change, contest and response. As the chapter unfolds, a series of concepts are identified that can be used to analyse environmental issues. These analytical concepts provide a means of helping to answer another key question in this book: how can we make sense of environmental issues?

The three subsequent chapters focus on extinction, a field of study in its own right in that loss of species and habitats has been occurring since life began, but it is the loss brought about by human action or inaction that is a matter of concern. The depletion of biological diversity is one of the most pressing environmental issues of our time and it is linked to some of the other major issues: global climate change, genetic modification, economic globalization, and deforestation. Extinction, then, allows us to answer a third key question: why are environmental issues so pressing? Extinction is also used here as a means of showing how analytical concepts can be applied to help us make sense of the environment.

Each chapter takes a set of concepts and applies them to aspects of extinction. Time and space are the concepts used in Chapter Two, which investigates the rates of extinction of species. It shows how species evolve over time and how quickly they can be eliminated. It demonstrates the importance of space in terms of the extent, availability and proximity of habitats to the survival of species. In particular, it shows how uneven pressures in different parts of the world, in conjunction with habitat destruction, have accelerated extinction in certain regions. We see how environmental change, in this case species evolution and extinction, is both a dynamic (temporal) and uneven (spatial) process.

Chapter Three focuses on the analytical concepts of values, power and action, taking as its subject the extinction of passenger pigeons in the USA and the threat of extinction to tigers in India. The examples show how values are important in defining the way we interpret the environment. They are often contested and, as a result, can be translated into interests that we pursue by exerting power. The consequences can be disastrous, as in the case of the extinction of passenger pigeons, or they may lead on to attempts at conservation, as is the case with tigers. Nevertheless, we see how the contests over environmental issues are likely to result in uneven development as one set of interests triumphs over others.

The conflict between conservation and extinction is compounded by the problems of risk and uncertainty, which are the analytical concepts introduced in Chapter Four. Here the subject is the monarch butterfly, with its fascinating and complex pattern of migration between the USA and Mexico. There are risks of hazards interfering with this pattern, possibly threatening the survival of the species. The problem of calculating risk is complicated by the many uncertainties of data and knowledge that are features of studies of environmental processes.

Uncertainty is an integral part of living and of scientific inquiry. Where possible we might want to reduce uncertainty. In other cases we may need to learn to live with it. Both of these responses need to be considered if we are to gain a better understanding of environments. Put another way, increasing awareness and knowledge is a key to understanding environmental issues and is the fundamental aim of this book and indeed the whole series. One way of trying to achieve this is to take an interdisciplinary approach – to show how individual ways of studying environments can together contribute to an integrated approach. So, each chapter here includes contributions from different disciplines.

Understanding Environmental Issues is, in many ways, a beginning. It is the first book of the series and, as such, it is an introduction to environment as a subject of concern and interest. However, it is also intended to be used as a resource, identifying the themes and concepts that, we believe, can be applied to the study of any environment or environmental issue. The book presents – perhaps for the first time – a coherent structure on which to base our approach to the study of environment and thereby provides a means to make our studies both accessible and transferable. As we proceed we shall begin to move from an intuitive recognition of the importance of environment in our daily lives to a deeper understanding of why environmental questions are so pressing.

Chapter One

Introducing environmental issues: the environment of an estuary

Andrew Blowers and Sandy Smith

Contents

'Between the mouths of the Blackwater and the Colne, on the east coast of Essex, lies an extensive marshy tract veined and freckled in every part with water. At high tides the appearance is that of a vast surface of Sargasso weed floating on the sea, with rents and patches of shining water traversing and dappling it in all directions. The creeks, some of considerable length and breadth, extend many miles inland, and are arteries whence branches out a fibrous tissue of smaller channels, flushed with water twice in twenty-four hours.

At noontides, and especially at the equinoxes, the sea asserts its royalty over this vast region.'

Revd Sabine Baring-Gould, *Mehalah*, 1983 (first published 1880)

1   Introduction: an estuary – where the elements meet

We begin with an estuary, a place where sea, land and sky meet. We have chosen a particular estuary: the Blackwater estuary on the Essex coast in eastern England. Although the Blackwater has its own unique characteristics, it is used here as a setting, a device for approaching the study of environments. Like any other estuary, the Blackwater brings together a diverse range of processes, elements and issues that constitute the environment. It offers us a way into thinking about how to approach the study of environments and why environmental questions are so pressing. By drawing on a variety of examples from the estuary we intend to convey how broad the study of environments can be; rather than focusing on a particular issue, this chapter is an introduction to the study of environments.

Imagine you are standing on the shore looking south out across the mouth of the Blackwater estuary (viewpoint in Figure 1.1). What do you see? Perhaps, at first, you take in the general scene, the setting. In front is the broad sweep of the river, which stretches from far upstream to where it first reaches the tide at Maldon and merges with the North Sea. Beyond, on the opposite shore, is the low-lying Dengie peninsula (Figure 1.2). Around you are the beach, saltmarsh and clay that make up the landscape surrounding the estuary. And above is the vast East Anglian sky, its importance in the scene emphasized by the flat country and sea all around (Figure 1.2). Here then, we have the three fundamental elements – using the term 'element' in the traditional sense of earth, water and air – the components that comprise the natural surroundings or environment, the **bio-** biosphere **sphere** (that part of the Earth capable of supporting life).

Figure 1.1 View over the Blackwater estuary towards the Dengie peninsula.

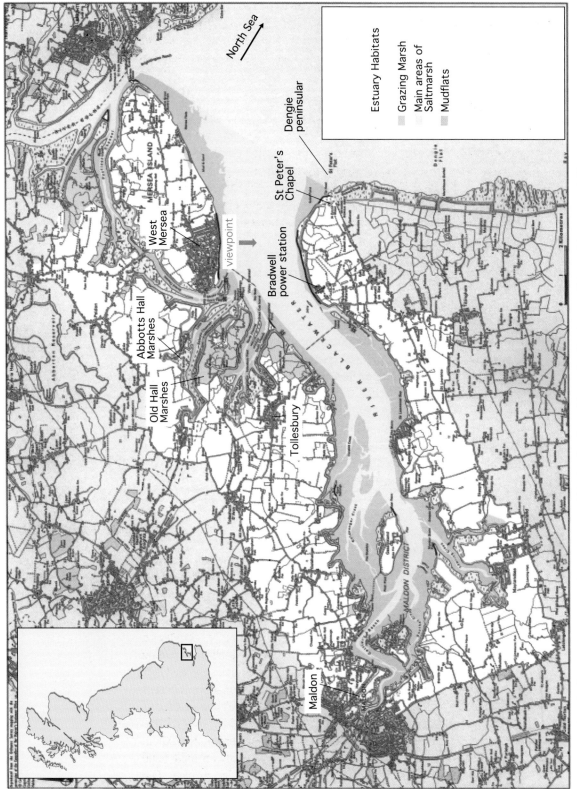

Figure 1.2 The Blackwater estuary, on the east coast of England, adapted from the Blackwater Estuary Management Plan.

Though the elements may be viewed as separate, they are also interlinked through various social and natural processes. Water, for example, which is present not only in the estuary but also on and in the land and as water vapour and droplets in the atmosphere, does not just stay in any one of these three elements but can also move between them (Box 1.1).

Box 1.1 The hydrological cycle

Water in the estuary can transfer to the atmosphere by evaporation from the water surface. Water in the atmosphere can fall on either the land or estuary as precipitation. From the land, rivers can move water to the estuary, and water on land can move through the land as groundwater and also evaporate from the surface or from the pores in leaves of vegetation to the atmosphere (a process called **transpiration**). This cyclic movement of water is called the **hydrological cycle**, or water cycle (Figure 1.3).

transpiration

hydrological cycle

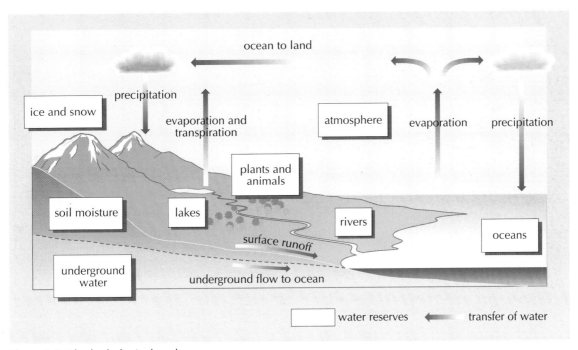

Figure 1.3 The hydrological cycle.

The scene before us is not static; it is one of ceaseless movement, not all of it evident to the human eye. There is the changing weather, sometimes creating spectacular storms that alter the shape of the coastline. There is the ebb and flow of the tides, revealing, at low tide, the gravel and muds of the foreshore. There is also the movement of life: the sea teems with fish and other sea creatures, including the oysters that are famous from this part of the world. The mudflats, too, support abundant life, and many of the species found here occur *only* in estuaries. It has been calculated that one square metre of mud can hold 1,200 worms, while 15,000 snails (*Hydrobia ulvae*) can graze on its surface. The saltmarsh forms a coastal **ecosystem** here (see Box 1.2) with a rich variety of

ecosystem

salt-tolerant plants (Figure 1.4). Above, there is the movement of flocks of birds along this important migratory highway (the so-called Eastern Atlantic Flyway).

Box 1.2 Ecosystems

The term ecosystem was first coined by Arthur Tansley (1935) to describe the interdependent groupings of plants, animals and other living organisms as well as inanimate components (such as water, soil minerals, etc.). All of these things use energy and process materials to produce the communities and landscapes that we see in the world around us. In ecology, the term has become associated with a particular approach to the study of the interactions among living organisms, an approach that concentrates on energy and material exchanges. However, it is also used more colloquially to refer to any area or grouping of organisms that can be regarded as functionally interdependent. An ecosystem in this sense can be of any size, from an isolated clump of mossy vegetation on a rock, to the whole thin layer around the world in which life occurs, the biosphere.

The estuary is never the same. It is always evolving, changing and recomposing. The environment is dynamic. It is a composition of rhythmic processes, of living movement and, from time to time, of sudden change. This feature of *changing environments* is the first major theme we shall be using in this book and the series as a whole to help you approach your studies. We shall return to this theme in terms of the estuary in a moment.

What other features compose the present scene? Across the estuary, near the tip of the Dengie peninsula, we can see the modest outline of the seventh-century St. Peter's Chapel. The Chapel is near the site of a Roman fort, testimony to the long period of human habitation in the area. About two miles to the west is the bulk of Bradwell nuclear power station, a twentieth-century edifice that provided electricity for 40 years (until 2002) and that will pose the risk of radioactivity leaking into the surrounding area for decades to come. On the water are the local fishing vessels, and in summer a myriad of boats and windsurfers are evidence of the recreational activities in the area. The beach huts fronting the sea indicate a seasonal holiday role. High above there are planes following the flight paths to London Heathrow and Stansted airports. Around the estuary the main occupation is farming, mainly arable (for this is in the cereal-growing east of the country) but also grazing, especially on the low-lying areas reclaimed from the marshes over the centuries. There are also the villages and small towns such as Maldon, Tollesbury and West Mersea together with smaller villages, with all the associated infrastructure of roads, housing, services and sewerage that make up the built environment (see Figure 1.2).

These activities and buildings reflect the different ways in which the environment might be valued by humans. The estuary provides resources that are fished and farmed. It also provides a means of dealing with wastes and pollution: after treatment, sewage is discharged into the estuary. Some radioactive wastes from

Figure 1.4 Saltmarshes are home to a rich variety of salt-tolerant plants.

Bradwell nuclear power station are either stored on site or emitted to sea or air, and other wastes and some toxic materials are disposed of to landfills. The estuary is also an artery giving channels for movement, for the transport of goods and people. Beyond these uses the estuary possesses recreational values for its resident and visiting human populations. These are all *instrumental* ways of valuing the environment: as a resource or use that is a means to an end. There may also be *non-instrumental values*, where environments are valued as ends in themselves, for their beauty, their rarity or their cultural qualities. These ideas are developed in Chapter Three, but here it is worth noting that environments can be valued for a variety of reasons. For example, the Blackwater is regarded as 'one of the most important habitats for wildlife in the UK' (Blackwater Estuary Management Plan in Maldon D.C. and Colchester B.C. 1996, p.1) and of international significance for the coastal ecology of mudflats and saltmarshes. (A **habitat** is the physical and biological environment in which an organism lives.)   habitat
The Blackwater's biological diversity has instrumental value as a resource for both humans and non-humans but it also has non-instrumental value in that it is conserved for its own sake. We shall return to the concept of biological diversity later in this chapter and it is considered in more detail in Chapter Two. For the present we shall note that biological diversity is more usually referred to as **biodiversity** (in its most basic sense, biodiversity is measured by the number and   biodiversity
variety of species in an area).

The values placed on the environment may give rise to *contest*, the second major theme of this book and series. This theme can help us understand why environmental questions are so pressing. Contests over values may result in conflicts between interests. Within the Blackwater estuary there are many different human interests each of which puts its own particular value on the environment: fishermen and farmers will value the environment as a source of livelihood; industrialists will value its capacity for absorbing pollutants or waste;

developers will focus on the market value of the location; wildfowlers prize its potential for sport; birdwatchers or ramblers will value the environment for its own sake, as an end in itself. These differing values often result in conflict between economic and non-economic interests. For example, around the Blackwater estuary the economic value of land for farming has led to the building of sea defences, the draining of saltmarshes and erosion of the marshes on the seaward side (Figure 1.5). This creates conflict between those interested in economic development (farming) and those interested in conservation (of an important habitat). The estuary provides many other examples of contested values leading to conflicts between interests, for example: conflict over fishing rights; conflict over the public's right of access to the shore; and conflict over proposals for urban or industrial development. In deciding how they value the environment and then pursuing their interests, different groups will not only come into conflict with each other but may have profound effects on non-human species and habitats.

Figure 1.5 With rising sea levels, areas of saltmarsh inevitably become eroded.

Each interest group will have access to various forms of political power, which they can use to promote their interests. For example, industrialists and developers have economic resources in the form of investment capital and employment; those with conservation interests are able to use the power of influence through the media or the support of local and national groups to exert pressure on decision makers. We shall develop these points later in this chapter and in Chapter Three. For the moment we simply wish to note that studying environments can also be associated with the idea of *contest* and *conflict*.

Returning to our viewpoint over the estuary, what other features are evident? Everywhere we can see signs of the effort to protect land from the incursion of the sea. Out to sea there is both dredging of channels and recharging of the foreshore with sand and gravel dredged from elsewhere (Figure 1.6). The aim is to create

Figure 1.6 Recharging protects the coast by creating offshore barriers.

Figure 1.7 Reinforced sea walls.

barriers to the erosive power of the waves on the shoreline. All around the estuary are earthen sea walls, often dating back to the Middle Ages, enabling grazing and later arable land to be reclaimed from the saltmarsh. In some places, especially close to human settlements, the walls have been reinforced (Figure 1.7). Elsewhere, in a few places walls have been deliberately breached to enable the re-creation of saltmarsh (see Figure 1.33). Saltmarsh dissipates the energy of waves and thereby provides a form of natural defence against the sea. On a coast along which sea levels are rising and that, from time to time, experiences destructive winds and tides, protection by reinforcing walls is only afforded to urban settlements and valuable structures, such as Bradwell nuclear power station. Elsewhere, the effort to resist the sea is regarded as both too costly and, in the long run, futile. The sea level rise predicted as a result of climatic change merely reinforces the need to provide 'softer' forms of protection (such as saltmarsh rather than sea walls) in the hope that creation of new marshlands will compensate for those that are lost through erosion.

All around this coast are examples of efforts to protect or enhance the environment. There are nature reserves, country parks and protected habitats, and the whole coastal fringe is designated as an area of scientific interest requiring special protection (Figure 1.8). There is also evidence of the need to manage the environment to ensure, so far as possible, compatibility between competing interests. Built development is prevented along the shoreline and restricted to existing settlements; caravan parks are carefully sited; and on the water the mooring and speed of boats are regulated. The environment of the estuary is the subject of regulation by agencies, planning by local authorities, and management by partnership and voluntary schemes. Its specific environmental qualities, such as its habitats and bird life, are defended by national and local

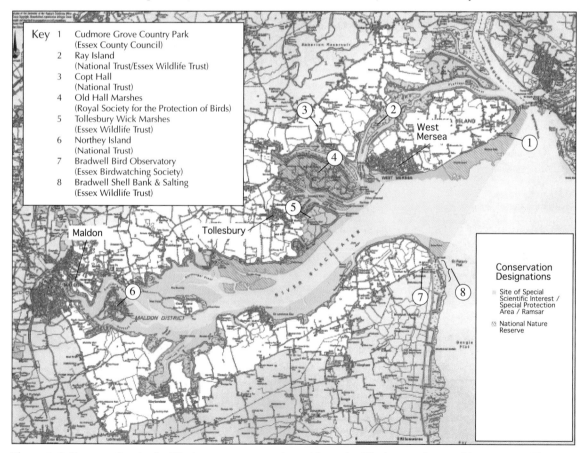

Key 1 Cudmore Grove Country Park
(Essex County Council)
2 Ray Island
(National Trust/Essex Wildlife Trust)
3 Copt Hall
(National Trust)
4 Old Hall Marshes
(Royal Society for the Protection of Birds)
5 Tollesbury Wick Marshes
(Essex Wildlife Trust)
6 Northey Island
(National Trust)
7 Bradwell Bird Observatory
(Essex Birdwatching Society)
8 Bradwell Shell Bank & Salting
(Essex Wildlife Trust)

Conservation Designations

Site of Special Scientific Interest / Special Protection Area / Ramsar

National Nature Reserve

Figure 1.8 Conservation in the Blackwater estuary, adapted from the Blackwater Estuary Management Plan.

environmental pressure groups and sometimes protected by legislation. From time to time, when a particular threat is perceived, local action groups may organize protests (such as those over plans for a nuclear waste site at Bradwell during the 1980s). The point we are making here is that environments evoke *responses* in the form of human action that is aimed at preventing, mitigating or managing change and conflict. In turn, environments respond to human action and this may lead to an alteration in human responses, and so on in a continuing mutual adaptation of human and natural processes.

This notion of *responses* is the third theme for this book and series. Response also suggests the idea of responsibility, not only for the present but for the future also. This is the kernel of the idea of **sustainable development**, endorsed at the United Nations Earth Summit at Rio in 1992 and which has since become a philosophical basis of much environmental policy making. In its most famous formulation sustainable development has been defined as development that 'meets the needs of the present without compromising the ability of future generations to meet their own needs' (World Commission, 1987, p.43). This definition combines the requirement of *sustainability* of the environment with th need for *development*, thus satisfying both 'the basic needs of all and extending to all the opportunity to satisfy their aspirations for a better life' (World Commission, 1987, p.44). Here, there is a recognition that environmental change both now and in the future will have different impacts on people and places and non-humans and habitats, reflecting their relative vulnerability. For example, people in comparatively affluent areas are better able to defend themselves (at least for a while) against rises in sea level than people in poorer low-lying areas (such as parts of Bangladesh) whose very survival may be threatened.

sustainable development

Sustainable development has been adopted as a fundamental principle of environmental policy making at all levels. For example, the idea has been translated (in the Blackwater Estuary Management Plan) into the specific context of the Blackwater estuary in the following terms:

> To promote the sustainable use of the estuary so that it can yield the greatest benefit to the present population whilst maintaining the potential for the estuary to meet the needs and aspirations of future generations, in a manner compatible with the maintenance of the natural properties of the estuary and its value for wildlife.
>
> (Maldon D.C. and Colchester B.C., 1996, p.8)

The marshlands and the animal and plant life they support are especially vulnerable to changes in climate and sea level, making sustainable development hard to achieve. As we shall see in later chapters, the aspiration of sustainable development in other contexts is difficult to translate into specific forms of action. We introduce it here to note the point that environmental responses combine both a reaction to contemporary problems and needs, and a conception of the kind of environment that should be bequeathed to the future.

Activity 1.1

How far, do you think, sea walls offer a form of defence against rising sea levels that meets the World Commission's definition of sustainable development? In what respects might the breaking of sea walls offer a sustainable response?

Comment

Sea walls, in so far as they protect settlements and farmlands, may meet the needs of the present. However, the continuing costs of maintaining them and the fact that in the long run they may prove ineffective against sea level change indicate that they may compromise the ability of future generations to meet their own needs. On the other hand, softer forms of sea defence, allowing the incursion of the sea, may destroy habitats and farmlands and, at times of tidal surges, leave settlements vulnerable. In the longer run they may prove more sustainable by enabling the re-creation of marshland habitats for future generations. Of course, in the longer term, no defence may be possible and all measures may prove unsustainable.

Now let us pause and consider what we have learned so far.

○   List the three major themes that we have introduced in this section.

●   Environmental change, environmental contest and conflict, environmental response.

The three themes form the basis of the following sections of this chapter and are also one of the ways in which this book and the series as a whole is structured. They also help us to interpret our first key question: how we can approach the study of environments? At this point we can make the following statements:

- Environments are never static but dynamic, driven by interacting physical and social processes. They change in ways that affect people and all other forms of life.
- Environments are open to a range of contested interpretations, values and understandings that sometimes lead to conflict between opposing human interests and between humans and non-humans.
- Environments are responsive to changes brought about by a variety of interacting processes (natural and human). In their turn, non-humans and humans respond to environmental change in a variety of ways. Environmental responses are uneven and environmental deterioration is more likely to affect vulnerable areas and populations.

In what follows we shall take, in turn, the themes of change, contest and response as ways of approaching the study of environments. We shall keep our estuary in mind but use it to explore some key analytical concepts that will reappear throughout this book and accompanying series. As you read on you should note the concepts that you think can be used to study environments. As we proceed we shall also begin to find answers to our second key question: why are environmental questions are so pressing?

2   Changing environments: the contemporary world in context

The movement and change we can observe in the estuary is created by the rhythms of tide, current, weather and living creatures. Over time we may note some significant changes as a mudflat is overspread by gravel, a saltmarsh is eroded (up to two metres a year in some places), or a beach is destroyed or a new one created by tidal forces. The shape of the estuary and the demarcation between land and sea is always shifting. Over longer periods of time the changes have been profound. What we are looking at today would have been different a few centuries or a thousand years ago and unrecognizable if we go back the millions of years that are expressed in terms of geological time.

2.1   Changes in historic time: the calamitous fourteenth century

There is a tendency to regard environmental changes in our own times as particularly significant in terms of the apparent acceleration of change (in large part due to human influence). New development and infrastructures impose greater pressures on environments and everywhere the noise of traffic, of powerboats and of aircraft intrudes upon the tranquillity. Around the estuary we may note the changes in the landscape brought about by intensification of farming, resulting in loss of hedgerows, trees, field boundaries and wildlife. Compare Figure 1.2 with the eighteenth-century map in Figure 1.9 to see the scale and impact of change. There are fears that we are living in an age in which human activity is causing devastating changes to ecosystems and an uncontrollable and possibly irreversible change in the world's climatic systems. Climate change has become a major environmental anxiety of the present age, and some believe we are witness to a transforming moment in terms of the rapidity and significance of environmental changes wrought by human actions.

If we go back, say, to the Middle Ages in Europe, we encounter both contrasts and similarities with our own epoch. Certainly, human occupation appeared less intrusive in those days, yet the time was also a period of widespread and cataclysmic environmental change. During the eleventh and twelfth centuries there was the so-called Little Optimum climate, a relatively warm period corresponding to a time of wealth and prosperity in the High Middle Ages. By the late thirteenth century the climate was deteriorating and, as the fourteenth century dawned, crop failures occurred as the 'Little Ice Age' set in and polar and alpine glaciers advanced, the Baltic froze over and the Caspian Sea rose (Figure 1.14). The Blackwater, at this time, would have occupied a slightly narrower and deeper channel than today. Barbara Tuchman has described the

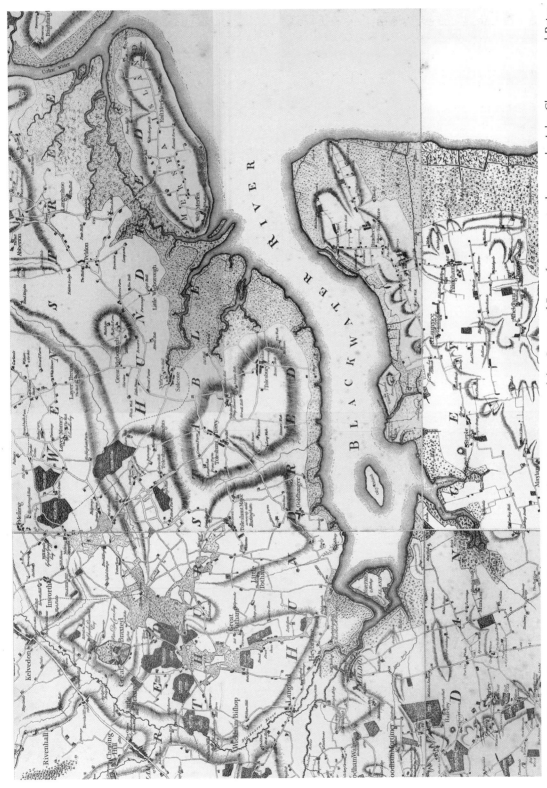

Figure 1.9 The Blackwater estuary in the late eighteenth century: a map of the County of Essex from an actual survey by John Chapman and Peter Andre, 1777. The illustration has been compiled from eighteenth-century maps in the Essex Record Office.

period as the 'calamitous' fourteenth century as environmental woes continued: 'In 1315, after rains so incessant that they were compared to the Biblical flood, crops failed all over Europe, and famine, the dark horseman of the Apocalypse, became familiar to all' (Tuchman, 1995, p.24). During 1319–1321 there was the great sheep murrain (anthrax), an infectious disease, herald of an even greater disaster that befell the human population in 1348 with the arrival of bubonic plague (commonly known as the Black Death) carried by fleas and spread by rats and travellers. Brought into southern Europe by sailors trading through the Black Sea with Asia, this virulent disease quickly spread across the continent killing over one third of the population in its first occurrence and substantial numbers during its resurgence in 1361 (Figure 1.10).

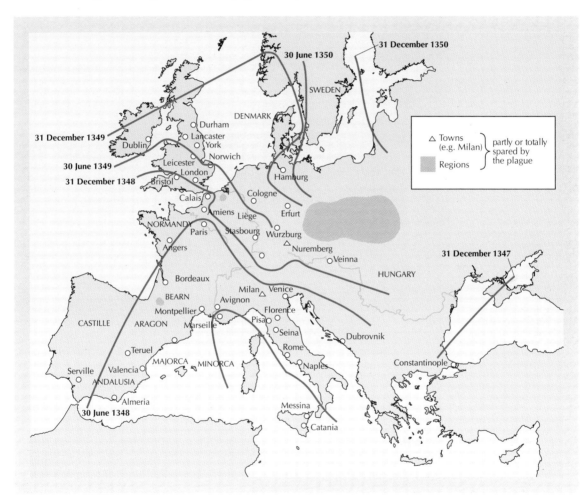

Figure 1.10 The spread of the Black Death: dates on the red lines show how far the first occurrence of the Black Death spread in just three years.
Source: Ziegler, 1997.

Given these calamities, visited upon the population for no apparent reason, it was little wonder that people believed it was the end of the world. Such an eschatological pronouncement has echoes in the doom-laden commentary of some pessimistic observers in our own times. Whether caused by the wrath of God as commonly supposed in the fourteenth century or the wilfulness of human actions as argued today, the sense of catastrophe is similar. At that time, a century and a half before the explorations charted the geography of the planet with which we are familiar today, the geographical knowledge of Europeans was limited so that the plague appeared to be a universal occurrence, a pestilence whose inexorable spread neither seas nor frontiers could resist. Although some accounts indicate that the plague had its biggest impact in areas where poor living conditions enabled it to spread, it affected all places and all social classes (Figure 1.11). There were no effective cures and, having winnowed the population, the plague eventually subsided only to recur from time to time with gradually diminishing impact.

Figure 1.11 The impact of the Black Death: plague scene from an Italian fourteenth-century manuscript.

The Black Death suggests that global environmental hazards are not confined to our own age but have occurred in earlier times. The simple point we wish to make here is that, in terms of perceptions of risk and uncertainty, the present era is not wholly unique. On the other hand, the Black Death, catastrophic as its impacts were, eventually disappeared. By contrast contemporary environmental changes, whose onset is slow but progressive, may possibly be irreversible, threatening the survival of human and non-human life alike.

2.2   Changes in geological time and space

If you had stood 18,000 BP (years before present) on the site of today's estuary you would not see an estuary at all; most of the area would be land. The sea would not be anywhere near here. You could walk across to what we now know as the Netherlands, as most of the North Sea and the English Channel would be dry land (Figure 1.12). You would also feel a bit chilly, as there would be the edge of an ice sheet about a hundred kilometres away to the north, and it would be advisable to keep alert for mammoths or other large mammals (Figure 1.13). These changes in the appearance of the area are related to natural climatic change on the Earth that is still continuing (Box 1.3).

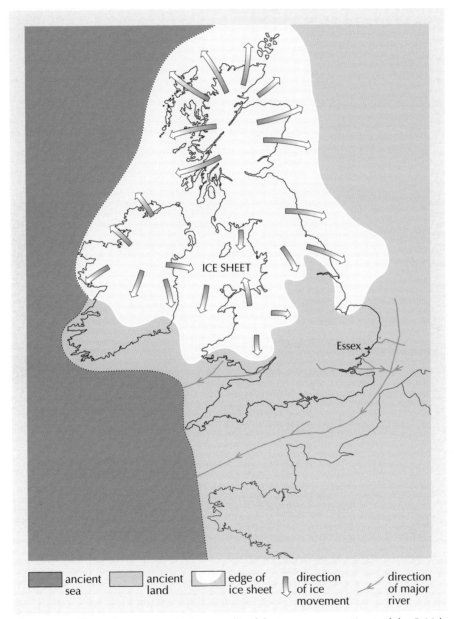

Figure 1.12 The *palaeogeography* (geographical features at a past time) of the British Isles 18,000 BP.
Source: Lucy, 1999.

Box 1.3 Ice Ages

Ice Ages are long-term colder periods of the Earth's history. At the moment we are in an Ice Age called the Quaternary Ice Age, which has lasted for about the last two million years. Ice Ages are characterized by many fluctuations in temperature (Figure 1.14a). We are in one of the relatively warm periods of this Ice Age, correctly called an 'interglacial Stage'. The cooler periods are called 'glacial Stages'. The present interglacial Stage has lasted since about 11,500 BP (Figure 1.14b). The 'Little Optimum' and the 'Little Ice Age' were small temperature fluctuations that occurred during the present interglacial Stage (Figure 1.14c).

During glacial Stages, the global sea level falls, for two reasons:

1 The polar ice sheets increase in size, removing water from the oceans and causing a fall in sea level.

2 The water remaining in the oceans contracts in volume as it cools, also causing a fall in sea level.

At the end of a glacial Stage, increasing temperatures will melt the ice sheets, returning water to the oceans, and the ocean water will expand. Both processes will cause a sea level rise.

Figure 1.13 A reconstructed view from the same point as in Figure 1.2 (towards the Dengie peninsula) as it would have appeared 18,000 years ago, during the summer. As an ice sheet lay just to the north of us, we would have needed warm clothing. The Blackwater (background) would have been a river not an estuary, flowing in summer but frozen in winter. Land would have extended across to the continent; the North Sea would not have existed, and the Blackwater would have flowed to the nearest sea which would have been to the south-west of England.

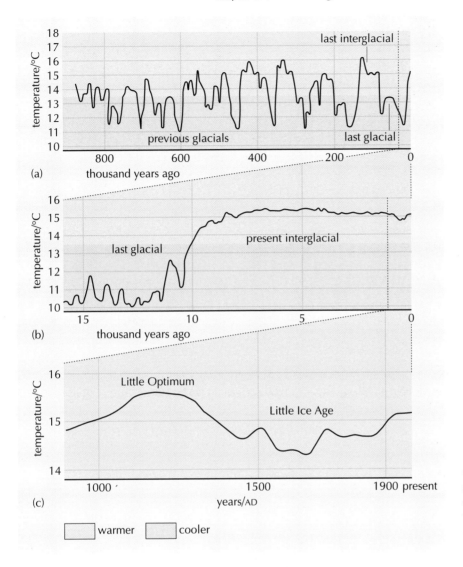

Figure 1.14 The estimated global mean surface temperature over the past 800,000 years.
Source: adapted from Houghton et al., 1990.

Activity 1.2

Considering the part of Figure 1.14 between 500,000 BP and the present:

(a) What is the general temperature change between a glacial Stage minimum temperature and an interglacial Stage maximum temperature?

(b) How often do glacial and interglacial Stages occur?

For the answers, go to the end of this chapter.

The maximum fall of sea level during the last glacial Stage occurred at about 18,000 BP, with levels going down to 120 metres below the present sea level and Britain joined by land to continental Europe (Figure 1.12). The switch to the

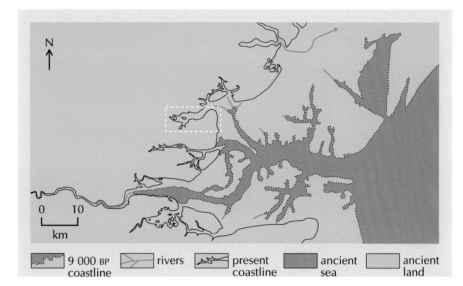

Figure 1.15 The coastline of Essex 9,000 BP. *Source*: Hunter, 1999.

present interglacial Stage occurred at about 11,500 BP accompanied by rising sea levels. By about 9,000 BP the North Sea and the River Blackwater (which existed before the glacial Stage) were re-established, but sea level was still lower than at present and the coastline was further to the east (Figure 1.15). Rising sea levels and the sinking of south-eastern England caused a further change in the palaeogeography, so that by 5,000 BP the coastline had moved nearer to its present position (Figure 1.16).

What about further in the past? The pattern of oscillating temperatures and changing sea levels continued throughout the Quaternary Ice Age. During an earlier glacial Stage in about 450,000 BP (the maximum extent of glaciation) an ice sheet reached south almost to the Blackwater area (Figure 1.17). The ice sheet diverted the river called the 'proto-Thames' – that later became the River Thames – further south from its older course, which ran across East Anglia to join the sea on the Essex coast near Clacton (Figure 1.17a). This proto-Thames deposited

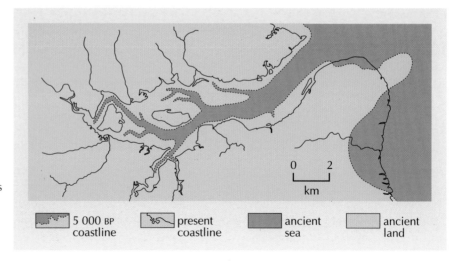

Figure 1.16 The Blackwater Estuary 5,000 BP. The location is indicated by a broken white line in Figure 1.15. *Source*: Hunter, 1999.

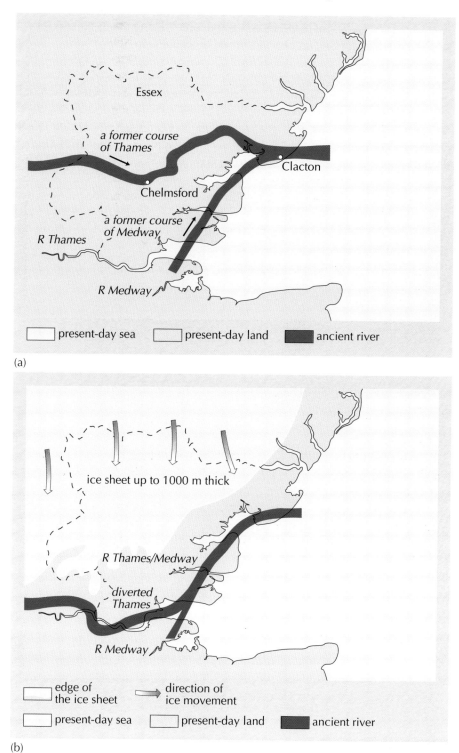

(a)

(b)

Figure 1.17 (a) The course of the proto-Thames and proto-Medway about 500,000 BP, before the arrival of the ice sheet in the area. (b) The maximum extent of the Quaternary ice sheet in Essex about 450,000 BP, and the course of the Rivers Thames and Medway after diversion by the ice sheet.
Source: Lucy, 1999.

river sands and gravels in the Blackwater area. After this, at about 400,000 BP, the River Blackwater formed its own valley and drainage pattern.

But what about earlier than this? Evidence of the environmental conditions of older times comes from a study of the rocks themselves. The rocks around the Blackwater estuary (other than the proto-Thames sands and gravels) are about 50 Ma (million years) old. This rock formation is called the London Clay and is composed mainly of clays and sands that were deposited under the sea, so the sea must have covered at least part of England at this time. Essex was just one of the areas in south-east England that was covered by sea (Figure 1.18). We also know from fossils in the rocks that it was much warmer. The land bordering the sea had dense subtropical forests with marginal palm and mangrove swamps. Fossil fruits of the palm nipa (*Nypa fruticans*) shown in Figure 1.19, now growing in Malaysia, can be found in the London Clay.

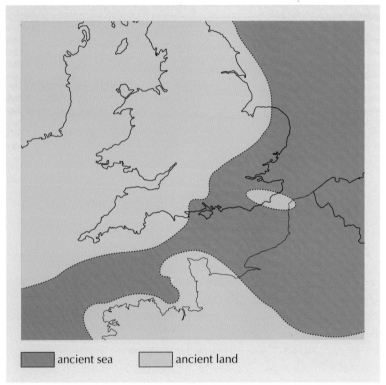

ancient sea   ancient land

Figure 1.18 Palaeogeography of southern Britain 50 Ma ago.
Source: Murray, 1992.

Not only was the Earth as a whole warmer 50 Ma ago (there was no Ice Age at the time) but Britain was in a different location on the Earth; it was closer to the equator at a latitude approximately 41°N, another reason why the British climate was warmer. Since then Britain has been moving slowly northwards, caused by a process called **plate tectonics (Morris and Turner, 2003)** in which areas of the Earth's surface (plates) can move relative to other areas. Today Essex is approximately 52°N, a movement of 11°N – about 1,200 km in 50 Ma.

plate tectonics

Rocks of 50 Ma may seem very ancient, but these rocks are youngsters in comparison with most of the rocks in Britain. The oldest British rocks are about 2,800 Ma old, and the age of the Earth itself is 4,600 Ma. Compare that with the time that modern humans have existed on the Earth, about 150,000 years. So, in geological time, natural changes have affected the environment of the Blackwater very significantly, without any human influence. Spatially, the sea level has oscillated up and down by over 100 metres in the last two million years, and Britain has moved over 1,000 km to the north in the last 50 Ma. Climate has also varied significantly, from subtropical 50 Ma ago, to a succession of cold and warm periods in the present Ice Age, with average temperature varying by about 5°C between the glacial and interglacial Stages. This gives a baseline for understanding present-day environmental change. Changes today are nothing like as great in amplitude as the natural environmental changes were in the past. What is of concern about the present changes, however, is the rapid *rate* at which they are occurring – perhaps unprecedented in the Earth's history.

Figure 1.19 The palm nipa (*Nypa fruticans*) growing along a river bank.

Summary

In this section we have focused on the idea of environmental change by putting contemporary changes in a historical and geological context. We have seen how our present anxieties about global changes have echoes in the past, though the circumstances are very different. And, by going back in time, we have seen how vastly different the environment was in terms of climate and landforms.

3   Contested environments: the use of resources

We return now to focus on more recent times. Looking back we are, perhaps, struck by the importance of physical processes in shaping the environment. Even in the Middle Ages there was a sense of the overwhelming power of natural forces, perhaps interpreted as divine intervention over human (and non-human) destiny. In the intervening period of scientific discovery and the impact of technology we have become accustomed (at least in the developed world) to the idea that we can, to some extent, control our environment. At the same time we are aware that there may be a limit to how much control we are permitted to exert and that, in certain cases, we cannot transgress those limits with impunity. In this section we shall once again use the estuary as an example, this time to consider environmental contests and conflicts and the implications for sustainable development. By using two examples – fishing and nuclear energy – we shall get some clues as to why environmental issues are so pressing.

3.1 Fishing: a sustainable resource?

In our first example we focus on conflict over **resources**, defined here in the sense of something that is useful or valuable for the human population. Fish provide a good example of a resource. Fishing is a source of food and income but there are concerns that current rates of extraction are too high in many areas throughout the world and that sustainability is threatened. However, restrictions on levels of fishing can produce hardship and conflict. Therefore, if we are to achieve sustainability in such fisheries we need to understand the environmental, ecological and economic conditions necessary for survival of fish stocks.

herring

sole

The Blackwater is an important fishing area; indeed, West Mersea has the largest inshore fishing fleet between Lowestoft in Suffolk and Brixham in Devon. The fish caught in the estuary and locally in the North Sea are mainly herring, sprat and Dover sole (also cod, whiting, plaice, dab, flounder, bass and grey mullet) and the area has an ancient, famous and still thriving oyster industry.

nutrient

Estuaries can be areas of abundant marine life, with shallow water suitable for marine plant growth and a good supply of **nutrients** (chemical elements necessary for life, even though present in only small amounts) carried from the land by the river. However, human uses of an estuary often conflict with the needs of marine life, destroying habitats and producing pollution from ports, power stations, agriculture and other development.

Fishing is a form of hunting. Hunting involves searching for and catching usually mobile animals that will not necessarily remain in one location, and to whom political boundaries have no meaning. The exception to hunting for mobile animals is fishing for shellfish, such as mussels and oysters, which live attached to the seabed (Figure 1.20).

The estuary is a spawning ground for fish (an area in which fish hatch) and a nursery ground (where fish grow as juveniles, before moving on to adult feeding areas elsewhere in the North Sea) and so is important to the North Sea fisheries. In the North Sea there has been over-fishing of some species, notably herring and cod, which led to the near collapse and closure of the herring fishery in 1977 and severe restrictions on both fisheries in the 1990s. Management of fisheries involves understanding what controls populations of fish (**Brandon and Smith, 2003**; **Drake and Freeland, 2003**) and the effect that fishing has on that population. Fishing for a limited stock of fish may cause conflict between different fishing boats and between different nations. To prevent overfishing, a system of licences, quotas, by-laws and other restrictions

Figure 1.20 Oyster fishermen on the estuary.

can be introduced. The North Sea is an international area with its fisheries mainly under EU control, and the need to conserve fish stocks has led to the allocation of catch quotas by the EU in an attempt to provide a catch that is sustainable.

○   In what sense is 'sustainable' being used here?

●   Ensuring that a balance is struck between exploiting and conserving the fishing stock.

The changes in the mature stock of North Sea herring are shown in Figure 1.21. Prior to 1950 the herring stock was near to five million tonnes. Herring in the North Sea were fished as adult fish for human consumption, and as juvenile stock in small-meshed nets catching a number of species for industrial use, such as fishmeal for animal feed. By 1977 the stock had fallen to 0.05 million tonnes. This was mainly due to overfishing the juvenile herring, but the stock was also affected by heavy fishing for other fish by bottom trawlers. This damaged the gravel sea-bottom substrate necessary for the successful spawning of herring. The stock was reduced to a non-sustainable level, that could not produce sufficient young for the herring stocks to increase or even be maintained at this low level, and in 1977 the fishery was closed – no fishing for herring was allowed.

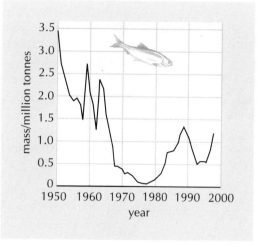

Figure 1.21 Annual trends in the mass of the mature stock of herring in the North Sea, 1950–1998. *Source*: CEFAS, 2002.

A slow recovery started, and the fishery was reopened in 1981. By 1990 the stock was much improved and was over one million tonnes. But the fishing rates for both adults and juveniles were high. The stock size began to fall rapidly to 0.5 million tonnes. In 1996 the total allowable catch of adult fish was halved mid-year, and restrictions were also put on the juvenile herring caught in the industrial fisheries. These very restrictive measures helped the stock, which slowly increased for the last part of the 1990s.

This example illustrates the threat posed by human activity to other forms of life. As we saw in Box 1.2, organisms within an ecosystem are interdependent and therefore the loss of one species will often have an impact upon many other species. The following chapters, which focus on extinction, will explore these issues in more detail. Here we wish to emphasize the inherent conflict between exploiting and conserving the resource. From the fishing industry's viewpoint the fish stocks represent a livelihood, one that may be threatened by overfishing so long as everyone has access to fish stocks and there is no incentive for anyone to withdraw. This is an example of the **tragedy of the commons**, an idea first put forward by William Foster Lloyd and developed by Garrett Hardin (1968). The tragedy results from a tendency to overuse

tragedy of the commons

Castree considers this idea further, particularly its political implications (**Castree, 2003**).

common resources to the point at which they become depleted, degraded, exhausted or extinct. In the case of fishing, so long as there is unrestricted access to a free resource there is no incentive for anyone to withdraw even if an area becomes overfished: the catch will reduce, but will still be better than nothing, though some fishermen may go out of business. Eventually there may come a point when there are no fish left.

There is a conflict here between the individual and the common interest. The common interest in a sustainable fishing resource can be protected by goodwill or, failing that, by collective and enforceable action. In the case of fishing this has meant the development of international regimes and agreements introducing quotas and other restrictions. In terms of sustainable development there are both natural elements (the ability to sustain life) and social elements (the ability to sustain livelihood). In an economic sense sustainable development requires that the resource is sufficiently maintained to provide the basis for continuing exploitation.

3.2 Nuclear energy: a changing issue

Bradwell power station (Figure 1.22) dominates the view across the Blackwater estuary. It was commissioned in 1962 as one of the first fully commercial nuclear power stations in Britain. Bradwell was selected for its remote site, its coastal location for cooling water and its relative proximity to the south-east of England where electricity supply was at the time deficient. Bradwell illustrates the shift in values that has brought environmental conservation into greater prominence. At the time the project was conceived, instrumental values, which emphasized production, technology and economic development, were dominant. Nuclear energy (our second example) was generally regarded as cheap, safe and clean, heralding a new technology bringing limitless supplies of electricity.

The public inquiry on the proposal for a power station held in 1955 lasted a mere five days (compare that to the 365 days needed for the inquiry into Sizewell B, on the nearby Suffolk coast, three decades later). The main environmental concerns raised were the aesthetic intrusion and the potential damage to the oyster breeding grounds. Little attention seems to have been paid to the risks to human health and to the environment from nuclear power generation and resulting radioactive wastes. The local newspaper reveals a singular confidence, even complacency, in the disdain for concerns about safety. The plant was 'inherently safe, it was only in deference to public opinion that the first two stations ... would be built away from large centres of population' (*Essex County Standard,* 2 February 1956). 'They in Bradwell were the guinea pigs for England and they might feel proud to be selected for this honour' (*Essex County Standard,* 11 May 1956). Indeed, the casual way nuclear risk was discussed is breathtaking to contemporary ears: 'Would an atomic explosion once every generation, or perhaps less, be acceptable as the price to pay for this new power?' (*Essex County*

Figure 1.22 Bradwell nuclear power station.

Standard, 25 October 1955). The problem of nuclear waste was dismissed: 'Methods of dealing with radioactive waste matter were so well known in all the countries of the world that there was no hazard attached to the disposal of them' (*Essex County Standard,* 18 November 1955).

Contrast this relative lack of concern about environmental hazards with the situation three decades later when Bradwell was identified in February 1986 as a possible site for the shallow disposal of some of the country's low- or intermediate-level of radioactive wastes (but nonetheless potentially harmful if not safely managed) (**Blowers and Elliott, 2003**). It was one of four sites in eastern England (the others were in Bedfordshire, Lincolnshire and Humberside) chosen by Nirex, the nuclear industry's waste disposal company, for comparative evaluation as to the sites' technical suitability. Already in Bedfordshire, which had been identified as a potential site more than two years earlier, a protest movement together with the Bedfordshire County Council had used a range of methods (petitions, media campaigns, lobbying, scientific expertise) to discredit the proposal. The Nirex proposal was met with similar responses at the other sites. At Bradwell, the only site where the nuclear industry was already established, the campaign was mobilized by Essex Against Nuclear Dumping (EAND) who established strong links with other similar campaigns, while the Essex County Council took a more distanced view, arguing that 'the case against Bradwell is materially different and more likely to succeed if presented separately' (Essex County Council, 1986).

It was, perhaps, the combination of intense local resistance and common purpose among the four communities that led to the success of the protests. The campaigns achieved national prominence, so much so that the protesters were dubbed by *The Times* in an excess of hyperbole to be 'Middle-class, middle-aged hooligans from middle-England' (19 August 1987). The campaigns culminated in round-the-clock blockades at each site (Figure 1.23), which for a time, prevented the contractors from beginning their geological investigations. Added to the unfavourable publicity, reports (including one from the House of Commons Environment Committee, 1986) cast doubt on both the technical necessity and the political viability of the shallow burial proposal. In May 1987, the Secretary of State for the Environment, Nicholas Ridley, conceded that dropping the proposals was 'the responsible course of action'.

Figure 1.23 Protesters pictured at Bradwell, 1986.

Public concern about the environmental problems associated with nuclear energy was again evident at Bradwell in 2000/2001 when proposals to incinerate low-level radioactive waste from the plant aroused local opposition. This was hardly a new development since radioactive waste had been incinerated there up to the late 1980s. Nevertheless, there was considerable scepticism and public meetings were held, ministers lobbied and local pressure groups formed to resist the proposal, which was eventually withdrawn.

Looking at these conflicts over radioactive waste at Bradwell two questions spring to mind. The first is why the opposition expressed by the local community increased over time.

At first sight the intensity of the opposition might seem surprising. After all, a shallow disposal site for radioactive wastes or an incinerator is much less of a threat than a reactor accident. In any case, much more dangerous high-level wastes are stored at Bradwell prior to their removal to Sellafield in Cumbria for reprocessing. The intensity of the protests may be explained in two ways. One is that radioactive waste disposal has a long-term presence affecting future generations, whereas nuclear power generation is relatively short-lived, with the waste being removed on closure and the building eventually decommissioned (though this may take several generations). The other reason is that concerns over environmental risks had been growing since nuclear power was first installed. The public confidence placed in new technologies such as nuclear energy had been largely supplanted by mistrust, anxiety and a concern for environmental conservation and the protection of the health of future generations. Over time there had been a decline in trust and confidence in experts: the secrecy of nuclear decision making, the awareness of the harmful effects of radioactivity and a series of accidents (Windscale, Cumbria, 1956; Three Mile Island, USA, 1979; and, most cataclysmic of all, Chernobyl, Ukraine, in 1986) had all contributed to the shift in values. The House of Commons Environment Committee commented that: 'Public anxiety is significant and deep-rooted' (1986, para.221). It is clear, then, that the increase in opposition reflects changes in values over time.

The second question is why the protesters were able to succeed. Again, success against a large and well established company may, intuitively, seem surprising. Obviously more information would be required to provide a full explanation. But from this short account it is clear that the protesters were able to deploy various resources of *power* through which to defeat the proposed repository. This extends our previous use of the term **resource** to include political and economic   resource
sources of power. These power resources available to the protesters included access to a basically friendly media, the ability to provide scientific information to counter the case for the repository, lobbying of significant decision makers, public meetings, petitions and so on. Above all, they were able to mobilize sufficient support within the local community to make their presence felt through peaceful protest, notably the blockade of the site. They also built alliances with other communities similarly engaged in conflicts over repository proposals. Taken together, these resources provided the community with power that they could translate into various forms of *action* to achieve the outcome they wanted. At the time, the power of the industry – based on its access to government, its provision of investment and employment and its scientific expertise – was unable to prevail. The secretive decision making and lack of consultation proved a powerful weapon in the hands of an opposition able to demonstrate that the process of site selection was unfair and premature and that the scientific case was open to doubt.

While the conflicts at Bradwell represent a defeat for the nuclear industry at the time, in other circumstances the outcome could be different. The nuclear industry's power may increase as it offers an alternative to fossil fuels as Britain's oil and gas supplies run out and the problem of greenhouse gas emissions becomes more serious. Decision making may become more open and participative thus embracing communities rather than provoking them. On the other hand, nuclear energy may remain too expensive and be considered to be too dangerous to justify further investment.

Summary

These two examples of fishing and nuclear power illustrate how values are contested and give rise to conflicts between interests. Both examples relate to sustainable development in the sense of ensuring needs can be met in the future. In the case of fishing we have a conflict over the use of environmental resources: whether to allow unrestricted exploitation or to try to conserve stocks for future human use through regulation. Conservation of stocks also helps to maintain the integrity of the ecosystem, thereby minimizing the effects of this conflict on non-human species. In the Bradwell case the impetus also comes from human needs but in a context of reducing the hazards to species (human and non-human) and habitats both now and in the future. The depletion or eventual exhaustion of fishing resources and the risk to health or survival of species and habitats from radioactive contamination are issues that demonstrate why environmental questions are so pressing. Not only are they pressing but they evoke *responses*, the issue to which we now turn.

4   Environmental responses: managing risk in conditions of uncertainty

Earlier we considered what is meant by 'environmental responses'. There are two aspects to this concept. One is the response made *by* the environment to processes of change, whether brought about by natural or human causes or a combination of both. The other is the response *to* environmental changes made by humans or non-humans. In this section we shall consider both of these aspects of response by focusing on an issue of particular significance in the Blackwater: how the environment of the estuary might be 'managed' in response to those changes that are having an impact on its biodiversity.

4.1   Change and response in the 'meadows of the sea'

wigeon

golden plover

golden samphire

sea barley

Midway along the northern shore of the Blackwater estuary lie the Old Hall Marshes on a peninsula three and a half miles long and lying between tidal creeks and mudflats (see Figures 1.2 and 1.24a). It is a wild, flat, remote area of mixed habitats of reedbed, open water, saltmarsh and improved and unimproved grassland. Ecologically, it constitutes coastal grazing marsh (Figure 1.24b) defined as 'periodically inundated pasture or meadow with ditches which maintain the water level, containing standing brackish or fresh water' (Essex Biodiversity Action Plan, 1999, p.145). These marshes, the so-called 'meadows of the sea', are vibrant with life. Old Hall Marshes contain 268 recorded plant species, among them the rare saltmarsh goosefoot (*Chenopodium chenopodioides*), sea barley (*Hordeum marinum*) and golden samphire (*Inula crithmoides*). There are over 24 mammals and about 1,000 species of invertebrates, and over 100 of them are rare. Fish and shellfish, including oysters, thrive in the creeks surrounding the marshes. Above all, these marshes provide breeding and feeding grounds for 240 species of birds, especially wildfowl (50 per cent of the total numbers in the Blackwater) like wigeon and teal, and waders (32 per cent) such as redshank, golden plover, dunlin and lapwing.

Old Hall Marshes are a crossroads, a fulcrum of bird migrations that link the local ecosystems to habitats thousands of miles away. Some species use the marshes as a stopover on the long journey south along the Eastern Atlantic Flyway to Africa, while others spend the winter here. Particularly prominent are

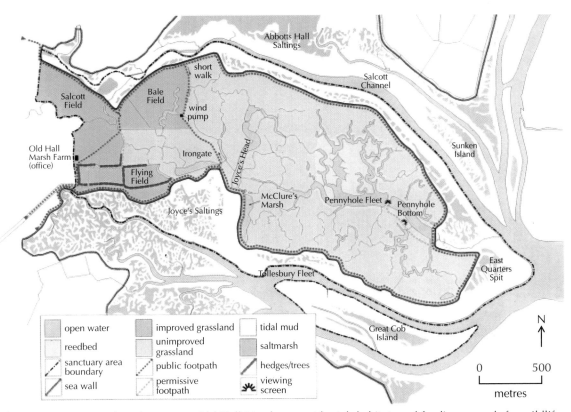

Figure 1.24a Protected environments: Old Hall Marshes provide rich habitats and feeding grounds for wildlife (map supplied by RSPB).

the dark-bellied brent geese that breed in the Taimyr Peninsula in the Arctic tundra of Siberia before wintering on the marshlands of the southern North Sea basin (Figures 1.25a and 1.25b). Old Hall Marshes support a peak of approximately 4,500 of these birds in winter, about two per cent of the total world population. In order to conserve the value of these marshes for habitats and wildlife they have been designated as a National Nature Reserve and with the rest of the estuary are protected in a variety of other ways (see Figure 1.8).

Figure 1.24b Grazing on the 'meadows of the sea'.

Conservation has become an important element in responses to natural and social changes in these marshlands. Old Hall Marshes are relatively recent, converted from saltmarsh to grazing marsh in a succession of reclamations that started in the sixteenth century and that are marked by counter-walls across the area. The purpose of reclamation was commercial, providing land for pasture for both sheep and cattle. Though remote, the marshes were just a two-tide journey by sea on a sailing barge to and from the London markets, so encouraging a profitable trade in animals reared or fattened on the marshes for

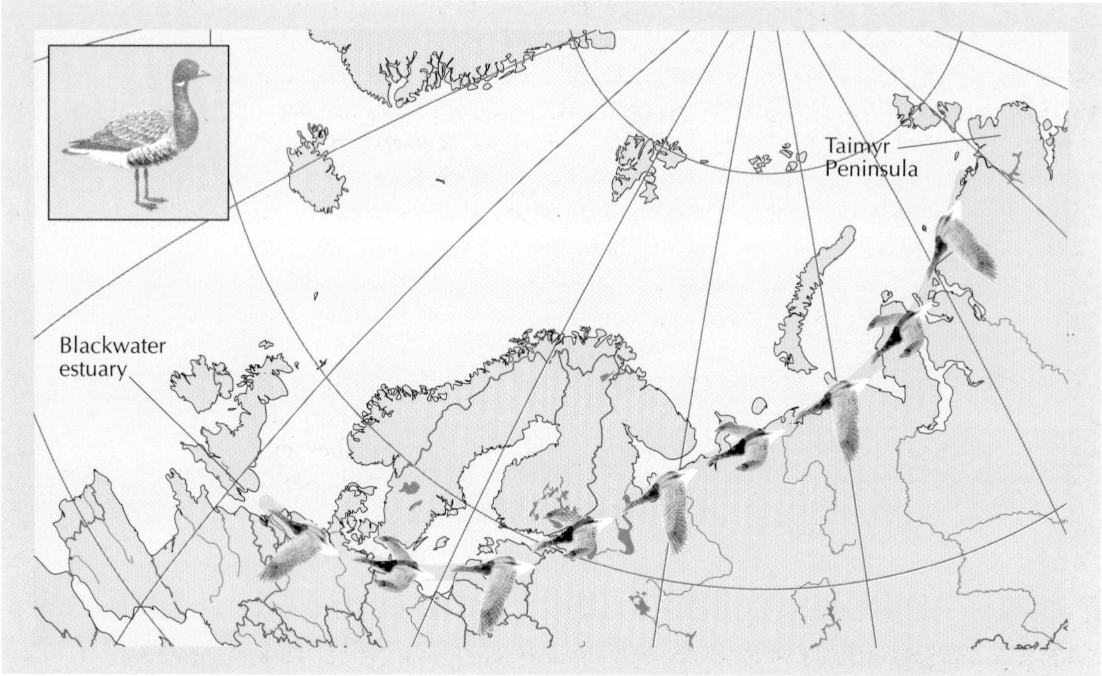

(a)

(b)

Figure 1.25 Brent geese: (a) their migration route to the over-wintering site in the Blackwater, and (b) in flight over the estuary.

meat or wool. The marshlands had other human uses, too: there are still traces of the decoy ponds used to ensnare wildfowl for fresh meat, and from the nineteenth century the marshlands attracted wildfowlers shooting birds for sport. Indeed the fact that this land supported large numbers of birds was a key reason for the survival of Old Hall Marshes while grazing marshes elsewhere were being ploughed up in the drive for food production during and after the Second World War. Since 1984, Old Hall Marshes have belonged to the Royal Society for the Protection of Birds, thus ensuring the conservation of the site as grazing land (cattle in summer, sheep in winter), grazing being essential for maintaining certain habitats and species.

Under the EU's Habitats Directive, coastal management must be sustainable; that is, there should be no net loss in terms of area of intertidal habitats. This poses a considerable challenge since in recent years both grazing marshes and saltmarshes along the Essex coast have declined considerably. Over two-thirds of grazing marsh was lost during the twentieth century, the biggest losses through conversion to arable land (Figure 1.26). But on the seaward side the protection of grazing marsh has contributed to the loss of saltmarsh. Since medieval times, enwalling has converted about 40,000 hectares of the Essex saltmarsh for agriculture leaving only 4,400 hectares, which is 10 per cent of the UK's total. This area has been eroding at the rate of two per cent per year (over one fifth of the total was lost during the last quarter of the twentieth century). The saltmarsh's gradual migration inland, as the sea level rises, is blocked by the sea wall, and the marsh is caught in a 'coastal squeeze' between the walls and the rising water. As Figure 1.27 shows, the gradual erosion causes a loss of important habitat as well as a reduction in an important form of coastal protection. With the onset of human-induced climate change the very survival of the coastal marshlands is at stake.

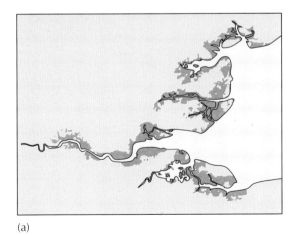 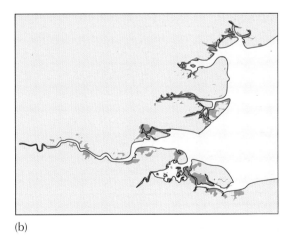

(a)   (b)

Figure 1.26 The decline in grazing marshes in Essex: (a) in the 1930s, and (b) in the 1980s. *Source*: RSPB.

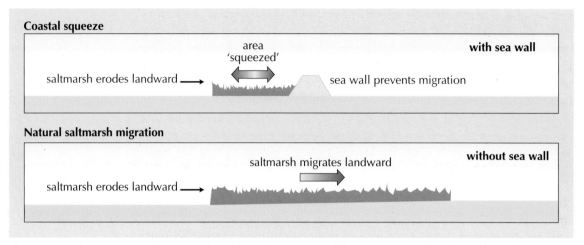

Coastal squeeze

saltmarsh erodes landward ⟶

area 'squeezed'

sea wall prevents migration

with sea wall

Natural saltmarsh migration

saltmarsh erodes landward ⟶

saltmarsh migrates landward

without sea wall

Figure 1.27 'Coastal squeeze' occurs when saltmarsh is prevented from migrating inland away from the rising sea level (adapted from Environment Agency).

4.2 Climate change: survival at stake

Despite efforts to define it, the boundary between land and sea is constantly changing. In the long run the combination of rising sea level, sinking land and possible major storms, such as the one that devastated the Essex coast in 1953 (Figure 1.28), indicates a battle that the sea must ultimately win.

Figure 1.28 The east coast floods: Jaywick Sands, Essex, February 1953.

The invisible threat of global warming is widely regarded as inevitable, provoked by the increased release of greenhouse gases, and likely to result in climatic extremes and significant rises in sea level within one or two generations. Predictions by the Intergovernmental Panel on Climate Change (IPCC, 2001) indicate that, by the end of this century, global sea level could have risen by as much as 0.88 metre as a result of global warming.

In addition to a global sea level rise, the Blackwater estuary is also affected by a local change in sea level. During the last glacial Stage, a thick layer of ice covered the whole of central and northern Britain. When this ice melted, and the weight on the land was reduced, the land in northern Britain began to rise (a process called 'glacial rebound'), tilting Britain and causing part of the south, including the Blackwater estuary, to sink. This caused a *relative rise* in sea level in this area, which is still continuing today, at a rate of about 0.15 metre a century. So for the Blackwater estuary, the rise in sea level is caused both by global warming and local glacial rebound.

Activity 1.3

From your reading of Section 2.2, state the two processes that would cause the global sea level to rise naturally at the end of a glacial Stage. For the answer, go to the end of this chapter.

There is much uncertainty about the scale, timing and impact of climate change. However, appreciation of the likelihood of significant changes in the foreseeable future has prompted recognition that response in some form is necessary if the possible consequences to human and other forms of life are to be avoided. We focus here on some of the uncertainties and possible effects of climate change, and on what the nature of the response might be.

For any estuary, such as the Blackwater, where much of the surrounding land area lies within a few metres of the sea (see the five metre contour around the Blackwater coast in Figure 1.29), any consistent rise in sea level could lead not only to destruction of the saltmarshes and widespread flooding, but inundation of agricultural and urban areas beyond. However, on a geological timescale, we have seen that sea level change of over 100 metres has occurred in the past. Part of the present-day concern is the *rate* of change, which is much faster now than at any time since the rapid warming and associated sea level rise at the end of the last glacial Stage at 11,500 BP.

In the Blackwater estuary, human response is in the form of various methods of coastal management. Defensive 'hard' sea walls are expensive, inflexible and difficult to maintain and are now used only to protect areas of urban development (Figure 1.7). Elsewhere 'soft' forms of coastal protection are used. For example, using brushwood polders to provide shelter from waves, using sunken Thames lighters as wavebreaks, or recharging the foreshore using dredged materials to break the force of the waves offshore (Figures 1.30, 1.31 and Figure 1.6).

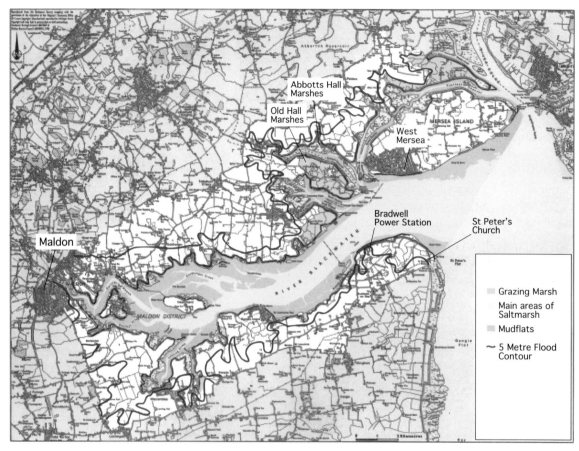

Figure 1.29 Land lying below the five metre mark at the Blackwater estuary, adapted from the Blackwater Estuary Management Plan.

In several parts of the estuary new saltmarsh is being created in various schemes described as 'managed retreat' or 'coastal realignment'. One such scheme, managed by the Essex Wildlife Trust, is at Abbotts Hall Farm, an area of arable farmland adjacent to Old Hall Marshes (Figure 1.32). The idea is to breach the three kilometre wall at five points, allowing ingress and dispersal of the sea, gradually transforming the low-lying meadows back into saltmarsh. On the rest of the farm, hedgerows, grassland margins, ponds and woodland are being created to encourage greater biodiversity. Under government-sponsored arable reversion schemes, financial incentives are offered to farmers willing to manage their land to enhance its value for conservation. Although coastal realignment has been successful in some places, such as Orplands (Figure 1.33), its impacts elsewhere may be unpredictable and uncertain.

Figure 1.30 Brushwood polders are a 'softer' defence than sea walls.

The present Blackwater coast has been created by a combination of natural and human responses to change. For several centuries sea walls have defined the border between reclaimed land and sea, while gradually, on the seaward side of the walls, the incremental loss of most of the saltmarsh has removed the first line of defence. Recently, the value of the saltmarsh biodiversity and the protection it affords has led to efforts to prevent further losses. At present, erosion far exceeds the creation of saltmarsh. Ultimately, if local sea level rise continues at its present rate of six millimetres per year, nothing can prevent the inundation of the Essex marshlands and areas well inland. Climate change will produce a range of problems affecting

Figure 1.31 Sunken lighters help to defend the coast against wave action.

all areas, causing loss of species or livelihoods. Drought-prone areas are likely to become larger and might occur just as they did in the past in the area that is now marshlands. Climate change could also lead to more *extreme* events, such as more frequent and heavier storms, and high sea levels (storm surges) which add to the flooding caused by sea level rise.

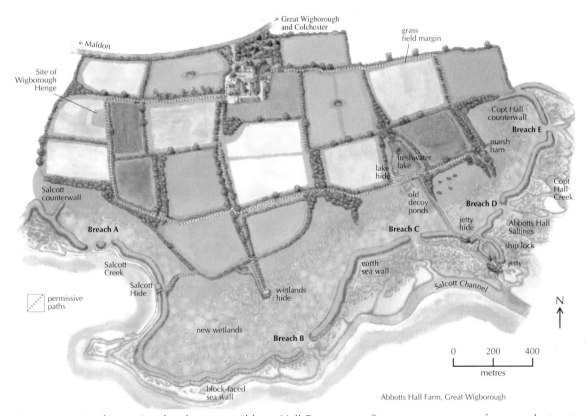

Figure 1.32 Map forecasting the changes at Abbotts Hall Farm over a five-year programme of managed retreat (from Essex Wildlife Trust).

Although we can adapt to the effects of climate change by retreating (e.g. from coastal areas), by accommodating them (e.g. growing flood-tolerant crops) and by protecting against them (e.g. building sea walls), we cannot prevent such effects. On a global scale, roughly one billion people live at or just above sea level, and cities such as London, New York, Tokyo, Bangkok and Shanghai could be submerged. Poorer countries are particularly vulnerable and have limited affordable options. Some, like the low-lying Pacific island of Tuvalu, face extinction and, in 2002, some of its population have already moved to New Zealand.

We can only slow the process of climate change by reducing greenhouse gas emissions, notably carbon dioxide emitted into the atmosphere from various sources such as fossil fuel power stations. However, before undertaking any drastic or expensive actions, governments need to know how good the predictions of the effects of global warming will be – how accurate are they? To examine this we need to look at predictions in more detail. The Intergovern-mental Panel on Climate Change (IPCC) report of January 2001 gave an estimate for a maximum global sea level rise of 0.88 metre by the end of the century. However, the minimum estimated change is only 0.09 metre. These predictions are made by making various assumptions, such as rates of greenhouse gas emissions and their re-absorption by, for example, vegetation (**Blackmore and Barratt, 2003**). You may wonder what is the use of a prediction that gives such a large range, 0.09–0.88 metre. This is a real problem, and such a range is an indication of the *uncertainty* of the result.

However, as uncertain as results like these are, they have to be used as a basis for policy decisions on global warming. If we waited until the end of the century to see whether sea level will rise by 0.09 metre or by 0.88 metre and then decided what to do about it, it would obviously be too late to stop it happening. It could be argued that it would be better to plan for sea level rise now. This is the basis of the precautionary principle   **precautionary principle** set out in the World Charter for Nature adopted by the UN General Assembly in 1982 – 'lack of scientific certainty should not be used as a reason for postponing measures to prevent suspected or threatened environmental damage'. The principle has subsequently been incorporated into various international conventions on the protection of the environment. The strength of the precautionary principle is that it may provide a basis for action when science is unable to give a clear answer. In the Blackwater estuary the rise in sea level is already sufficiently palpable for precautionary action to be taken.

Human responses to environmental change may be conceived in terms of reducing risk. We define risk as the probability of a hazard occurring at some future point in time or as a result of a particular action. An example of the former might be the risk of a coastal habitat being destroyed within a hundred years. The risk of saltmarsh loss through building a sea wall is an example of risk resulting from a particular action. Thus risk implies a probability of change, and therefore requires prediction of the likely outcomes of response. The confidence that can be placed in any prediction depends on the level of uncertainty about the processes responsible for change. As we have seen, there is considerable uncertainty about the future change of sea level. But in terms of our key question the issue is so pressing that we cannot afford to wait and see. In these circumstances a precautionary response in the hope that the impact may be mitigated is the best we may do.

Activity 1.4

Throughout this section we have discussed how the Blackwater environment has been 'managed' in response to changes in sea level. Try now to identify some of the human responses to change in the Blackwater, including those both beneficial and detrimental to the conservation of the estuary's biodiversity.

Comment

Responding to the incursion of the sea into residential and agricultural areas, humans have built solid sea walls and left undefended the saltmarsh on the seaward side of the walls. This has resulted in the progressive erosion of the saltmarshes, thereby threatening the species that depend on these areas for their livelihood. People have also created 'softer' forms of coastal defence, such as brushwood polders for breaking waves.

Coastal realignment, however, is a response that works with (rather than defends against) the constantly increasing sea levels. In breaching sea walls and allowing saltmarsh to migrate inland as the sea level rises, humans are working for the greater long-term benefit of Blackwater biodiversity. Humans have also introduced various types of designation to ensure the marshlands biodiversity is both conserved and enhanced (see Figure 1.8).

Summary

In this section we have once again used the Blackwater to identify a key theme for the chapter, this time the theme of environmental *response*. Responses involve the interaction of human and natural processes. For example, the rise in sea level experienced in the estuary is partly natural readjustment after the Ice Age and partly a response to greenhouse gas emissions caused by human activity. This rise has brought about two forms of response: human responses to change in the environment and responses made by the environment.

The first form, responses by humans to the impacts of environmental change, includes the building of sea walls to prevent the loss of productive land, and various measures of coastal protection and conservation management to prevent further loss of saltmarsh and its biodiversity.

Figure 1.33 Managed retreat at Orplands.

The second form of response, that made by the environment to changes, includes: coastal erosion as a response to sea level rise; extinction of species through loss of habitat; and loss of saltmarsh through coastal squeeze.

You may note that responses tend to have both negative and positive impacts. Thus, responses by and to the environment have resulted both in the loss of saltmarsh by flooding and embankment, and in its re-creation through managed retreat (Figure 1.33).

5   Conclusion: themes, concepts, questions

We have used the Blackwater estuary as a way of exploring how we can approach the study of environments. There are three structuring elements: core themes, core concepts and key questions.

All the books in this series adopt this approach.

5.1   Core themes

The second book in the series, *Changing Environments*, examines how processes acting over time and space have shaped our environments.

The first core theme is that of *change*: the recognition that environments are dynamic, composed of interrelated natural and social processes. We illustrated this feature by looking back over historical and geological time. Understanding how environments change enables us to place contemporary changes in context so that we can evaluate their significance.

The third book in the series is *Contested Environments*.

The second theme is that of *contest*: the notion that environments are subject to different understandings and that sometimes these lead to conflicts over the use that is made of environments. We have used the example of conflicts over fishing and nuclear energy to demonstrate this feature.

The fourth book in the series, *Environmental Responses*, will show how societies are both responsible for change and have to find mechanisms to deal with its impacts.

The third theme is that of *response*: the idea that the vulnerabilities inherent in environmental change evoke reactions in the form of adaptation or adjustment on the part of ecosystems (which include human communities). Response also suggests more proactive forms of human action intended to influence the direction, nature or pace of change. We examined this feature through the examples of estuary management and climate change.

These three themes are obviously interrelated. Environmental change influences and is influenced by environmental contests and conflicts which, in turn, inspire responses that attempt to deal with environmental change. We cannot understand everything at once, so we shall focus on each theme in turn while recognizing their inherent connections.

5.2 Key analytical concepts

You may have noticed that in this study of the Blackwater we have focused, at different points, on some specific concepts. For example, we have emphasized the importance of *time* in understanding environmental change and of *uncertainty* in relation to environmental responses. Earlier we suggested you might be able to identify some of the concepts we have been using as ways of approaching the study of environments. *Time* and *uncertainty* are two of them. Did you find any others?

Well, the concepts we have been using can be separated out into three groups.

Time and space	Values, power and action	Risk and uncertainty

These we are going to call *key analytical concepts* and they are the second structuring element we shall use. The use of key analytical concepts as a structuring device is found throughout the series, with each book focusing on a particular set of concepts. They will be used as devices to interpret environmental processes, issues and policies; as means to encourage interdisciplinary awareness; and as conceptual tools to enable you to understand the content of each chapter and to achieve its learning outcomes. At this point we are simply identifying the concepts; however, as with the core themes, the analytical concepts are only separated out as a matter of convenience for teaching purposes. They, too, are obviously interlinked, as should already be clear from the discussions in this chapter.

The use of key analytical concepts as a structuring device is found throughout the series, with each book focusing on a particular set of examples.

Taking each of the analytical concepts in turn, can you think of some of the ways they have been used in our exploration of the Blackwater estuary?

Time and space   *Time* has been quite explicitly identified, notably in the examples of the fourteenth century and geological time presented in Section 2. The idea was to show how minuscule is the period of human occupation of the Earth when compared to geological time. Such a timescale and our place within it is almost impossible to imagine. Box 1.4 explains this comparison in terms of an analogy we can grasp. In this context, the environmental changes now occurring may be compared with those in the historical past and with the cataclysms of the remote past. We can also realize that we occupy a tiny fragement of time within an interglacial stage, though the changes now occurring could have devastating consequences within a very short period of one or two generations.

Although we have focused on time as a linear concept (i.e. moving along from a starting to a finishing point), you may also have noted that time can also be conceived of in terms of rhythms, some that are repetitive like tides, and day and night or the seasons, others that are less predictable such as the timing of ice ages. Concepts of time matter when we are trying to understand environmental processes or make predictions.

Box 1.4 Humans on Earth

The Earth was formed, along with the rest of the solar system, an unimaginable 4,600 Ma ago. Modern humans (*Homo sapiens sapiens*), however, only evolved 150,000 BP. To put this into perspective, if we imagine the history of the Earth as one calendar year long, with the Earth forming on 1 January, modern humans would not appear until the last day, 31 December. Even then, we do not arrive until about 17 minutes before the end of the year.

Space has not been treated so explicitly as time. However, the importance of space is evident in the nature of an estuary as a physical entity. The estuary comprises various aspects of landscape, land use, ecosystems, activities and processes that provide it with its unique combination of characteristics and its identity. Although bounded, it is also connected to other places. The river links it upstream with other landscapes and localities and it flows out into the North Sea and ultimately into the great oceans, which cover nearly three-quarters of the surface of the Earth. Migrating birds, such as brent geese, spend the winter in the Blackwater estuary but then move on over vast distances to other parts of the world. Transport by sea, air and land enables flows of products and people to and from the area, flows that are constantly changing through complex interconnections. Although the Blackwater estuary defines an area, its connections elsewhere also define it as part of a set of global processes and interactions. Changes in time and space are also connected: we have seen how the size of the Blackwater estuary varies on a geological timescale with changes in sea level, and that it has also moved in a northerly direction. So, we may conceive of space in a variety of ways: in terms of *scale* (local to global); as *bounded* (in terms of natural and political/administrative boundaries) but also *transboundary* (as environmental processes cross political/administrative boundaries); as *distance* with places connected and separated by *spatial relationships* such as migration, transport and trade. *Locations* may also shift physically through the process of plate tectonics.

Values, power and action So far, we have identified two contending sets of values: those that are instrumental (directed at the use of the resources of an environment) and those that are non-instrumental (concerned with preserving an environment for its aesthetic and cultural significance). Of course, the two are not completely opposite, or even separate. For example, a concern to protect an environment may both reflect the wish to enjoy the amenity and also maintain property values. A sustainable fishing policy both conserves the environment and ensures a continuing supply of fish. As we have seen, values are often expressed in terms of interests. The various human interests may be compatible (as in the case of sustainable fishing) but they are often in conflict (as between the nuclear industry and local protesters).

The outcomes of conflicts will vary over time and space depending on the power available to the contestants. Their power may vary over time depending on prevailing values or economic circumstances. Thus, the nuclear industry possessed considerable and unopposed power during its early years but latterly has faced strong opposition as its economic position has declined and environmental groups have gained in strength as concern about risks has increased. Possession of power does not necessarily imply action, or at least not overt action. If you think back to the Bradwell example, the nuclear industry remained unchallenged for a long period. It was only when new facilities for dealing with the problem of radioactive waste were proposed during the 1980s that public concern grew (Figure 1.34) and the industry was forced to respond and defend its interests. The exercise of power influences the distribution of environmental resources. There is a pattern of uneven development as powerful communities are better able to defend, and perhaps enhance, the quality of their environment, sometimes at the expense of less powerful communities that may experience increasing environmental deterioration. For instance, if the disposal of nuclear waste is rejected in one location it must still be managed somewhere else. There is a rough correlation between wealth and environmental

Figure 1.34 Newspapers reporting the Bradwell protest, 1986.

quality. Again, the climate change example has illustrated the point. It is, in the first instance, the poorer parts of the world that are likely to suffer most from the predicted impacts – though, in the long run, the impacts are likely to be indiscriminate.

Risk and uncertainty The Blackwater estuary provides several examples of environmental risk. There are the risks of biodiversity loss, for example loss of fish, and the loss of saltmarshes and mudflats, and their associated species. There are the risks of environmental deterioration caused by pollution of the water and the atmosphere and from built development. There are also the longer-term risks of damage from radioactivity and climate change. Although we usually talk of the need to reduce risks, sometimes we must *take* risks: we have to gamble on an optimistic outcome to our actions (for example, that managed retreat will create more saltmarsh).

On the positive side, environments can improve as well as deteriorate: in the Blackwater area, water quality has improved; sewage treatment has enabled bathing beaches to achieve the necessary standards; levels of the toxin TBT (tributyltin) in the water are declining; wildlife is being conserved through estuary management; saltmarshes are being protected through managed retreat; speed limits are reducing the noise on the river; and sustainable practices in fishing and farming are being introduced.

However, we have seen in the cases of radioactive waste, biodiversity management and climate change that risks are difficult to predict. There is considerable uncertainty about the rate of environmental change, the likelihood of impacts occurring and the distribution of those impacts should they occur. A certain level of risk may be regarded as acceptable, but above this level efforts may be made to reduce the risk. Thus, in the case of radioactivity a common measurement is that the risk of developing a fatal cancer for a member of the general public should be no higher than one person in a million per year. In seeking to achieve such targets a range of assumptions must be made and prediction becomes more and more difficult the further ahead one looks. All we need to note at this point is that uncertainty is a common condition. In an area like climate change the scientific – let alone the social – variables are so many and complex that while there may be some consensus over general trends, the predictions of the timing, impact and consequences of changes are open to considerable debate.

5.3 Key questions

These are the third of the structuring elements of this book and the series as a whole. In this chapter we have tried to give some answers to the question: how we can approach the study of environments? We have also posed a second question: why are environmental questions so pressing? We suggest you consider this question now as you finish this chapter. It provides a useful bridge to the following chapters, which focus on the subject of *extinction* as a way of discussing the key analytical concepts and demonstrating the importance of environmental issues. As you think back over this chapter on the Blackwater estuary try to relate it to your own environment or one that is familiar to you. As you do so, try to use the themes, concepts and questions that we have introduced. By doing this you will begin to think analytically in a way that will help you to understand and enjoy the chapters that follow.

Different key questions will be posed in each book in the series.

References

Association of Essex Councils Steering Group (1999) *Essex Biodiversity Action Plan*, London, HMSO.

Baring-Gould, S. (1983) *Mehalah*, Woodbridge, The Boydell Press (first published 1880).

Blackmore, R. and Barratt, R. (2003) 'Dynamic atmosphere: changing climate and air quality' in Morris, R.M. et al. (eds).

Bingham, N., Blowers, A.T. and Belshaw, C.D. (eds) (2003) *Contested Environments*, Chichester, John Wiley & Sons/The Open University (Book 3 in this series).

Blowers, A.T. and Elliott, D.A. (2003) 'Power in the land: conflicts over energy and the environment' in Morris, R.M. et al. (eds).

Brandon, M.A. and Smith, S.G. (2003) 'Water' in Morris, R.M. et al. (eds).

Castree, N. (2003) 'Uneven development, globalization and environmental change' in Morris, R.M. et al. (eds).

CEFAS (2002) www.cefas.co.uk/fisheries/nsstocks.htm (accessed 6 February 2002).

Essex County Council (1986) *The Defence of Bradwell*, Colchester, Essex County Council.

Drake, M. and Freeland, J.R. (2003) 'Population change and environmental change' in Morris, R.M. (eds).

Hardin, G. (1968) 'The tragedy of the commons', *Science*, vol.162, pp.1243–8.

Houghton, J.T., Jenkins, G.J. and Ephraume, J.J. (eds) (1990) *Climate Change: The IPCC Scientific Assessment*, Cambridge, Cambridge University Press.

House of Commons Environment Committee (1986) *Radioactive Waste: Session 1985–6, First Report*, London, HMSO.

Hunter, J. (1999) *The Essex Landscape: A Study of its Form and History*, Chelmsford, Essex Record Office Publications.

Intergovernmental Panel on Climate Change (IPCC) (2001) *Climate Change 2001: The Scientific Basis. Contribution of Working Group 1 to the Third Assessment Report of the Intergovernmental Panel on Climatic Change*, Cambridge, Cambridge University Press, or http://www.ipcc.ch/pub/reports.htm

Lucy, G. (1999) *Essex Rock: A Look Beneath the Essex Landscape*, Saffron Walden, The Essex Rock and Mineral Society.

Maldon District Council and Colchester Borough Council (1996) *Blackwater Estuary Management Plan*, Maldon D. C. and Colchester B. C.

Morris, R.M. and Turner, C. (2003) 'Dynamic Earth: processes of change' in Morris, R.M. et al. (eds).

Morris, R.M., Freeland, J.R., Hinchliffe, S.J. and Smith, S.G. (eds) (2003) *Changing Environments*, Chichester, John Wiley & Sons/The Open University (Book 2 in this series).

Murray, J.W. (1992) 'Palaeogene and neogene' in Cope, J.C.W., Ingham, J.K. and Rawson, P.F. (eds) *Atlas of Palaeography and Lithofacies*, London, Geological Society.

Tansley, A.G. (1935) 'The use and abuse of vegetational concepts and terms', *Ecology*, vol.16, pp.284–307.

Tuchman, B. (1989) *A Distant Mirror: The Calamitous Fourteenth Century*, London, Macmillan.

World Commission on Environment and Development (1987) *Our Common Future*, Oxford, Oxford University Press.

Ziegler, P. (1997) *The Black Death*, Stroud, Sutton Publishing.

Answers to activities

Activity 1.2

(a) The glacial Stage minimum (global *mean* surface temperature) is about 11 °C, and the interglacial Stage maximum is about 16 °C, so the temperature change is about 5 °C.

(b) There have been about 4 glacial and interglacial Stages in the last 400,000 years, which suggests that they occur about every 100,000 years.

Activity 1.3

When glacial Stages end, and as the temperature rise of the next interglacial Stage progresses, global sea level will rise as (Section 2.2):

(a) The polar ice sheets melt, adding water to the oceans.

(b) The water in the oceans expands on warming.

Chapter Two

Are too many species going extinct? Environmental change in time and space

Joanna Freeland

Contents

1   Introduction: is extinction a cause for concern?

Extinction is defined as 'wiping out' or 'annihilation': if a certain type of plant or animal becomes extinct, it simply no longer exists. This may occur on a local or regional scale or, in the most extreme cases, on a global scale. In this chapter we will be focusing on the process of extinction, a theme that will provide us with some answers to the key question of how we make sense of environmental issues. Throughout this chapter we will also be continuing our discussion of why certain environmental issues are particularly pressing, another key question that was introduced in Chapter One.

The potential importance of extinction will start to become clearer if you think back to the estuary in Chapter One. As with all ecosystems, the estuarine ecosystem would not exist without the species that inhabit it. Furthermore, there is a fairly specific group of species that make an estuary (or any other ecosystem) 'work'. In other words, while species need ecosystems in which to live, it is also true that ecosystems cannot function without certain groups of species.

Should extinction be a concern to us? We recognize the link between species and ecosystems: as more species die, ecosystems also become threatened with extinction, and without ecosystems our planet as we know it would simply not exist. Or is this necessarily the case? It is hard to believe that life on Earth could exist without ecosystems, but perhaps there are so many species on Earth that the extinction of some will make very little difference. Do we really need to have approximately 2,500 species of ants, or 25,000 species of orchids? These are difficult questions to answer, and this difficulty leads us to the second reason for using extinction as the theme of this chapter.

The analytical concepts of time and space, which were introduced in Chapter One, can provide insight into a variety of issues regarding extinction. For example, one way in which we can determine whether the current extinction rates should give us cause for concern is to examine how extinction rates have varied through time – if the current rates have increased substantially over the past 200 years or so, perhaps we should start to think about why these increases may be happening. Alternatively, we can use the concept of space to look at extinction patterns around the world. If certain parts of the globe show noticeably higher extinction rates than other parts, we can begin looking for differences between these areas. Even a partial understanding of why extinction rates vary in time and space should improve our ability to predict what will happen to biodiversity over the twenty-first century.

One further point to note is that time and space can be used to analyse all sorts of environmental issues. As you read through the chapter, try to keep this in mind, and from time to time think of other issues that may benefit from similar analyses. Some suggestions of issues with which you may be familiar include climate change (**Blackmore and Barratt, 2003**), genetically modified organisms (**Bingham, 2003**) and radioactive waste (**Hinchliffe and Blowers, 2003**).

2   Have extinction rates varied over time?

When extinctions occur, they are most commonly documented as extinctions of *species*. This does not mean that species are the only concern in this context; for example, types of habitat and ecosystems can also go extinct. However, because habitats are complex and can be difficult to define, it is easier to measure the presence or absence of single species. As a result, species can be considered as a main 'unit' by which to measure extinction and extinction rates, and so we need to understand what the term 'species' represents.

2.1 What is a species?

All living things are thought to have evolved from a common ancestral species, and therefore all have varying levels of relatedness to one another. These relationships form the basis for dividing living things into a hierarchy of groupings that make up a commonly used system of **taxonomy**. The first of these groupings is kingdoms. There are various views on how the kingdoms should be divided; we shall adopt the classification that recognizes seven kingdoms: fungi, plants, animals, proctista (including algae, plankton and protozoa), bacteria, crenarchaeota and euryarchaeota (you may not have heard of the last two kingdoms, which are bacteria that live in harsh environmental conditions such as high salt concentrations or extreme temperatures). Within each kingdom, there is a series of categories in which individuals are placed according to their overall similarity to one another: phylum, class, order, family, genus and species. Table 2.1 contains examples showing how three different species are classified.

Taxonomy is the science of classifying and naming organisms. A taxon (plural = taxa) is a group of organisms within a particular classification.

Table 2.1   One way of classifying species: examples of fungi (mushroom), plants (meadow buttercup) and animals (human)

	mushroom	meadow buttercup	human
Kingdom	Fungi	Plantae	Animalia
Phylum	Basidiomycota	Angiospermaphyta	Chordata
Class	Hymenomycetes	Dicotyledoneae	Mammalia
Order	Agaricales	Ranales	Primates
Family	Agaricaceae	Ranunculacae	Hominidae
Genus	*Agaricus*	*Ranunculus*	*Homo*
Species	*bisporus*	*acris*	*sapiens*

Requirements for the different kingdoms are often fairly intuitive, for example we are generally confident about whether something is a plant or an animal. Phylum, class, order, family and genus are less straightforward, as they are artificial ways of classifying species into groups based on certain features that they have in common. The species, on the other hand, is a more tangible taxonomic unit

biological species concept

because, according to the **biological species concept**, it can be defined as a group whose members actually or potentially interbreed and produce viable offspring. Nevertheless, biologists are often unable to categorize species precisely. For example, the biological species concept is commonly violated by species that are asexual or by species that sometimes produce viable hybrids (a hybrid is the offspring of two different species). The difficulties of identifying species are compounded by the fact that in some cases two members of the same species may look very different, and in other cases two members of different species can look very similar (Figure 2.1). That said, most individuals can be assigned to species with a high level of confidence. Following the system devised by the eighteenth-century Swedish biologist Carl Linnaeus, each species is given a scientific name in Latin, made up of two parts. The first part of the name is that of the genus to which a species belongs (e.g. *Homo*), and is shared with closely related species. The second part of the name is that of the species itself. All humans are classified as a single species, *Homo sapiens*, and we have no living relatives thought similar enough to be placed in the same genus.

Although we refer here to hybrids as being the offspring of two different species, gardeners or farmers may be more familiar with the concept of hybrids as the offspring of two different varieties within the same species.

(a)

(b)

(c)

(d)

Figure 2.1 Appearances can be deceptive; they are not always useful for identifying species. This illustration shows two members of the same species: (a) female and (b) male ruff (*Philomachus pugnax*); and two members of different species: (c) Blyth's reed warbler (*Acrocephalus dumetorum*) and (d) marsh warbler (*Acrocephalus palustris*).

Nobody knows how many species exist on Earth today. About 1.82 million species have been given a scientific name, but this is a very incomplete sample of what is out there. In 1982, Terry Erwin of the Smithsonian Institute (Washington, DC, USA) published a report in which he proposed that estimates of biodiversity on Earth are seriously underestimated. He suggested that there might be 30 million species of insects alone. This conclusion was based largely on his finding that, in the tropical rainforest, insect species were often specific to individual trees. If any given tree in a tropical rainforest houses a number of unique insect species, it follows that there must be many millions more insect species than were previously believed. Estimates of the total number of species living today commonly range from 5 to 50 million, with some estimates as high as 100 million.

One reason why many species remain unidentified by humans is the high cost that would be involved in employing countless biologists to investigate in detail all areas of the globe. Species in remote areas, species with limited geographical distributions, and very small species (such as bacteria) are particularly likely to remain undetected. Furthermore, the number of species is just one way of measuring **biodiversity**, which refers broadly to how many different types of living things there are at any particular time or in any particular place. If we are to fully assess biodiversity we also need to take into account how much variation there is *within* species (**genetic diversity**; see also Box 2.1). Bearing in mind the uncertainty surrounding our estimates of biodiversity, we will use our understanding of species as a basis for examining extinction.

biodiversity

genetic diversity

Box 2.1 Biodiversity

Biodiversity is simply a contraction of the term 'biological diversity' and is a concept that embraces the variation of life on Earth. More specifically, we can break biodiversity down into concepts such as genetic diversity, species diversity and habitat diversity.

Genetic diversity refers to the variation of genes within a species. This encompasses genetic variation both *within* (Section 3.3) and *between* populations, with the latter referring to how genetically distinct populations are from one another.

A population is a collection of individuals belonging to the same species, living within a defined geographical area.

Species diversity is another way in which biodiversity can be calculated. This refers to all the different species (from all kingdoms) that exist within a region, and many assessments of species diversity take into account the number of individuals within each species in that particular region. Under this method of assessment, a pond that is home to 20 species, with each species represented by 10–20 individuals, is much less diverse than a pond that is home to 20 species with each species represented by 200–500 individuals.

A third measure of biodiversity is *habitat diversity*. This encompasses the variety of habitat types found within a region. Under this classification, a forest that includes a stream and some areas with reduced tree canopy (perhaps in a cleared area) is more diverse than a forest whose trees are evenly spaced and which lacks a stream or other physical disruption that would support distinct habitats.

2.2 Current patterns of extinction

How many plants and animals will become extinct this year? This is a difficult question to answer precisely but, to give you some idea, consider the following

The IUCN Red List is a list of species that are in danger of extinction. It is updated every few years and is based on several criteria for each species, including total population size, recent changes in population size, and geographical distribution (see Chapter Four for a more detailed discussion).

statistics. The 'Red List' of the International Union for Conservation of Nature and Natural Resources (IUCN, also known as the World Conservation Union), compiled in 2000, states that a total of 11,046 species of plants and animals are threatened with extinction in the near future, in almost all cases as a result of human activities. Of known species, this includes 24 per cent of mammals, 12 per cent of birds, 25 per cent of reptiles, 20 per cent of amphibians and 30 per cent of fishes. BirdLife International lists 1,186 birds as being threatened worldwide, up by 75 since 1994, with 99 per cent of these threatened as a result of human activity. A total of 182 of these bird species are described as critical, which means that they have only an estimated 50 per cent chance of surviving either the next 10 years or 3 generations, whichever is shorter. Over the last 500 years, 128 bird species have become extinct, most of these (103) since 1800 (see Figure 2.2). Since only a small proportion of reptiles, amphibians,

(a)

(b)

(c)

(d)

Figure 2.2 Examples of species that have recently become extinct: (a) ivory-billed woodpecker (*Campephilus principalis*), extinct 1990; (b) blue pike (*Stizostedion vitreum glaucum*), extinct 1983; (c) thylacine or Tasmanian wolf (*Thylacinus cynocephalus*), extinct 1936; and (d) (details of flowers of) sophora toromiro tree (*Sophoro toromiro*), extinct in the 1950s (in the wild).

and fishes have been assessed, the percentage of threatened species could be much higher.

The list compiled by the IUCN leaves out many types of animals, most notably invertebrates (animals without backbones), which comprise the majority of species found today (see Figures 2.3 and 2.4). This is largely because, although more numerous, these species are less well known than larger, more charismatic groups such as birds and mammals (Figure 2.5). Similarly, the fate of many plant species is not known. A total of 5,611 threatened plants is listed by the IUCN, but as the status of only approximately 4 per cent of the world's described plants has been evaluated, the true percentage of threatened plant species must be much higher. For example, 16 per cent of conifers (the most comprehensively assessed plant group) is known to be threatened, and therefore the overall proportion of declining plants may be similar to that of some animals.

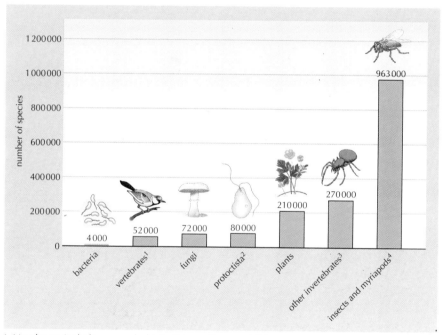

1 Vertebrates include mammals, birds, fish, reptiles, amphibians
2 Protoctista are mainly micro-organisms
3 Invertebrates that are not in any other category of the figure, including nematodes, crustaceans, molluscs, arachnids, etc.
4 Myriapods include centipedes and millipedes

Figure 2.3 This bar chart shows seven different groups; a total of 1.75 million species have now been described. It is evident from this figure that insects and myriapods are by far the largest group, with approximately 963,000 currently described.
Source: Data are from from Groombridge and Jenkins, 2000.

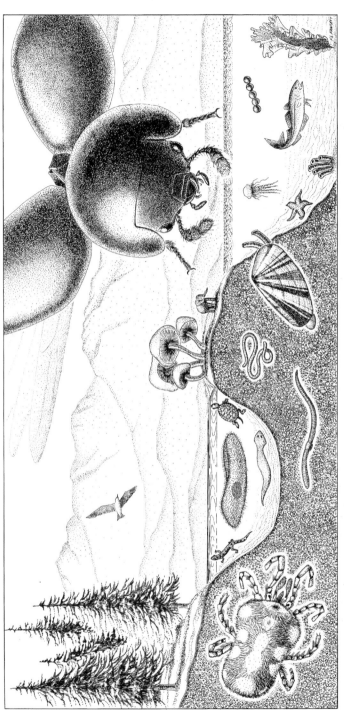

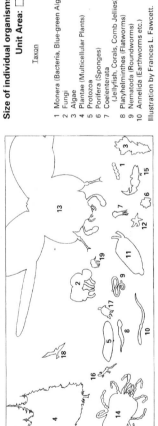

**Size of individual organisms represents number of described species in major taxon.
Unit Area: ☐ = approximately 1,000 described species.**

Taxon	No. of Described Species		Taxon	No. of Described Species
1 Monera (Bacteria, Blue-green Algae)	4,760		11 Mollusca (Mollusks)	50,000
2 Fungi	46,983		12 Echinodermata (Starfish etc.)	6,100
3 Algae	26,900		13 Insecta	751,000
4 Plantae (Multicellular Plants)	248,428		14 Non-insect Arthropoda (Mites, Spiders, Crustaceans etc.)	123-161
5 Protozoa	30,800		15 Pisces (Fish)	19,056
6 Porifera (Sponges)	5,000		16 Amphibia (Amphibians)	4,184
7 Coelenterata (Jellyfish, Corals, Comb Jellies)	9,000		17 Reptilia (Reptiles)	6,300
8 Platyhelminthes (Flatworms)	12,200		18 Aves (Birds)	9,040
9 Nematoda (Roundworms)	12,000		19 Mammalia (Mammals)	4,000
10 Annelida (Earthworms etc.)	12,000			

Illustration by Frances L. Fawcett.   From O.D. Wheeler.   1990.   Ann. Entomol. Soc. Am. 83:1031-1047.

Figure 2.4 Scaled illustration of the numbers of species within different taxonomic groups, as known to date. The more species there are within a group, the larger is that group's image. Proportions are based on numbers of species that have currently been classified: the giant black beetle, representing all insects, for example, reflects just under 800,000 species (current estimates indicate there may be 5–10 million or more!). Illustration by Frances L. Fawcett.
Source: Wheeler, 1990.

(a)

(b)

(c)

(d)

Figure 2.5 Examples of currently endangered species: (a) giant panda (*Ailuropda melanoleuca*); (b) whooping crane (*Grus americana*); (c) purple cat's paw mussel (*Epioblasma obliquata obliquata*); and (d) (female) hazel pot beetle (*Cryptocephalus coryli*). Which of these is more likely to headline a campaign that is trying to obtain public support for conservation?

At first glance these numbers appear to be rather alarming. But how do we know that this is not simply the way of the world? Perhaps in any given century, for example, up to one quarter of all species may go extinct. This could represent a steady loss of some species that is counterbalanced by the appearance of new species. Alternatively, extinction may occur in a cyclical or recurrent manner, an example of which would be short bursts of high extinction levels every 500 years or so. If either of these scenarios (continuous or cyclical extinction) is the case, then the current extinction rates may be perfectly natural and beyond our control. In order to examine these possibilities we will now explore what is known about extinction rates over time. Although we are at this point primarily concerned with the analytical concept of time, it is worth bearing in mind that the concept of uncertainty, as introduced in Chapter One (and to be further developed in Chapter Four) is another theme running through our comparison of current and historical extinction rates.

2.3 Extinction rates over time

In this section we see how important the fossil record is in helping us to piece together extinction rates over time – and yet how incomplete a picture it is able to provide for us.

Geological time is divided into two intervals: the Cryptozoic eon (4,600 Ma ago to 545 Ma ago), and the Phanerozoic eon (545 ma ago to present). The Phanerozoic eon is divided into eras, which are further subdivided into periods of unequal length (see **Morris and Turner, 2003**).

The fossil record

In order to assess historical patterns of extinction, we must first think back to Chapter One, where we learned that the Earth is approximately 4,600 Ma (million years) old. The earliest traces of life discovered by humans are in rocks that are 3,850 Ma old, although life probably arose on Earth before that time. Vertebrates did not appear until much later (Figure 2.6).

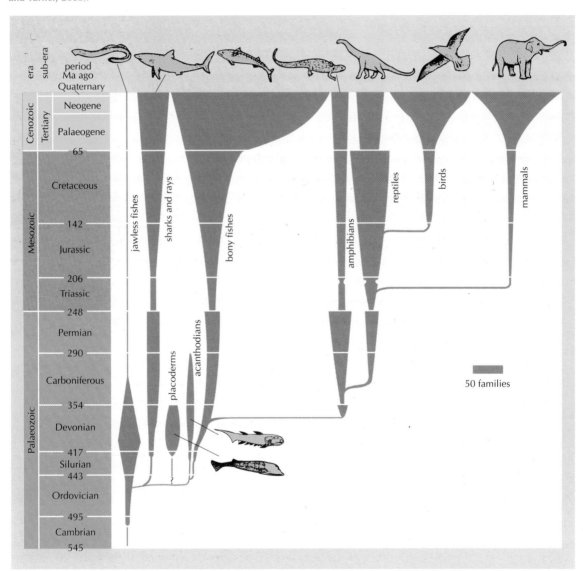

Figure 2.6 Geological timescale of the Phanerozoic showing when various vertebrate groups were believed to have diversified.

Activity 2.1

To get some idea of when different groups of organisms appeared, take a look at Figure 2.6 and answer the following questions:

(a) Which of today's groups of vertebrates were in existence by the end of the Triassic Period?

(b) When were (i) jawless fishes, and (ii) amphibians at their highest diversity in terms of numbers of families?

(c) In which geological period did birds first appear?

(d) What does this figure tell us about changes in biodiversity over time?

Answers to these questions are given at the end of this chapter.

Figure 2.7 A Jurassic fossilized ammonite (*Dactylioceras commune*), a cephalopod mollusc.

When we discuss rates, we are interested in how numbers change over time. For example, if you walked across France, covering 10 miles each day, you would be walking at a rate of 10 miles per day. Similarly, extinction rates refer to the numbers of species that have gone extinct over a certain unit of time, such as a year or a century or a millennium. Estimates of extinction rates over life's 3.5 billion year history come from the fossil record. Fossils are the remains or impressions of organisms that have usually been petrified (changed into stony substance) while embedded in rock (Figure 2.7). They represent our only knowledge of species that went extinct before humans began keeping such records.

How complete is the fossil record?

The fossils that have been discovered and described by palaeontologists (people who study life from the geologic past) represent more than one quarter of a million species, but these are believed to make up only a very small fraction – possibly only 1 per cent – of all the species that have ever existed. In scientific studies our information on a particular subject may be incomplete, and in these cases conclusions will be based on a certain amount of educated guesswork. With respect to the fossil record, part of the incompleteness is due to the fact that species that lived in fairly small areas, or those that survived for only relatively short periods of time, are less likely to leave their mark in the fossil record than are widespread, long-lived species. Furthermore, some species are much more likely to become fossilized than others. Small animals with hard skeletons including marine creatures such as molluscs (e.g. mussels and clams) and corals feature prominently in the fossil record of the last 600 million years. Soft-bodied organisms (such as jellyfish, illustrated in Figure 2.8, or slugs) seldom leave remains that become fossilized. Larger organisms are often devoured by scavengers and then decomposed to the point that little or nothing remains, and it is for this reason that dinosaurs make up a very small proportion of fossils.

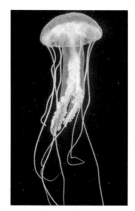

Figure 2.8 This jellyfish (*Pelagia noctiluca*) is unlikely to be preserved in the fossil record.

Hard-bodied insects have been readily preserved in the fossil record. For example, dragonflies more than 300 million years old have been found in the fossil record, making this one of the oldest groups of organism still existing today. (See illustration on chapter contents page.)

What can the fossil record tell us about extinction?

Despite the fact that most species have not been preserved in the fossil record, we can draw a number of conclusions about changes in biodiversity over time. Layers of rocks can be dated, and therefore the ages of fossils can be estimated based on the rocks in which they are embedded. Although life may have emerged as early as 4,000 Ma ago, relatively complex organisms (an **organism** is any living being, and complex organisms are those consisting of more than a single cell) did not appear until 610 Ma ago. From what we know, biodiversity during the first 3,500 Ma or so of life was low, with a relatively small number of species that generally survived for extremely long periods of time (hundreds of millions of years). Approximately 600 million years ago, a diversity of species appeared in the fossil record. Since this time, there has apparently been a steady increase in global levels of biodiversity, although it is worth remembering that this apparent increase is based on fossils from only a small proportion of species and therefore does not necessarily represent the full picture (see Figure 2.9).

organism

A cell, which is microscopic, is the basic structural unit of which all living things (except viruses) are composed.

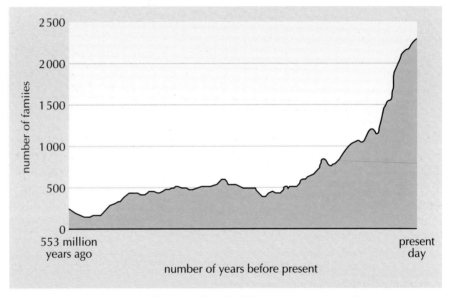

Figure 2.9 Total numbers of families identified from the past 553 million years. *Source*: Data from M. Benton at University of Bristol Earth Sciences website, http://palaeo.gly.bris.ac.uk/frwhole/Fr2.families.

Despite this overall increase in biodiversity levels over the past 600 million years, we know from the fossil record that up to 99 per cent of species that once lived are now extinct. Evidently the appearance of new species has, to some extent, counterbalanced extinction. In order to understand how biodiversity remains high despite ongoing extinction, we will first look at how such a large number (99 per cent) of species became extinct over time. These extinctions can be broadly placed into one of two categories: the background extinction rate and mass extinctions. A closer examination of these two categories will show us how the analytical concept of time allows us to use past events as clues that will help us to understand current events.

Background extinction rate

The background extinction rate refers to how often extinction occurs in the normal scheme of things, for example if species were killed by disease, predators, lack of food or a combination of such events. The background extinction rate has been variable over time, with an estimated average rate of 25 per cent of species going extinct each 1 million year period. Overall, this pattern of fluctuating rates of extinction is believed to account for up to 96 per cent of all extinctions that have occurred to date, an estimate based on species that are present in the fossil record and were shown to have disappeared during times other than mass extinctions (Raup, 1994). Note once again that there is a measure of uncertainty in this estimate – if we aren't sure how many species existed, our estimates of extinction rates are likely to be somewhat erroneous.

Mass extinctions

The other category of extinctions is mass extinctions, which account for the approximately 4 per cent of extinctions over time that were not included in the background estimates. Mass extinctions can be set apart from the normal background extinction rate because they refer to monumental events in which an estimated 75 per cent or more of all species existing at the time became extinct. At these times, extinctions led to a temporary decrease in overall biodiversity.

We can recognize in the fossil record at least five periods of mass extinction during the past 600 Ma. Once again, when reading the percentages listed below, bear in mind that these are calculated from a fossil record that does not represent all species, and are therefore only our best estimates. The five periods of mass extinctions are:

- Late Ordovician (440 Ma ago), in which approximately 85 per cent of species became extinct
- Late Devonian (365 Ma ago), in which approximately 85 per cent of species became extinct
- Late Permian (245 Ma ago), in which more than 90 per cent of species became extinct
- Late Triassic (208 Ma ago), in which approximately 75 per cent of species became extinct
- Late Cretaceous (65 Ma ago), in which approximately 85 per cent of all species became extinct.

Activity 2.2

Turn back to Figure 2.6 and mark on it approximately when each of the five mass extinctions occurred.

Figure 2.10 Images of dinosaurs rarely fail to capture our imagination. This figure is a representation of a river floodplain in North America, pictured during the Upper Cretaceous. The diversity of plants includes conifers, ginkgos, cycads and ferns. *Tyrannosaurus rex*, a large carnivorous dinosaur, emerges from the forest to attack the duck-billed dinosaur, *Edmontosaurus*. Overhead fly *Pteranodons*, winged reptiles with seven metre wingspans.

None of these mass extinctions occurred overnight – each probably took between 0.5 and 3 Ma, and the beginning and end of each mass extinction cannot be clearly differentiated from the background extinction rate. Nevertheless, mass extinctions are identifiable as periods of unusually high loss of biodiversity.

The most famous mass extinction is the one that ended the Cretaceous period 65 Ma ago, well known for being the time when the dinosaurs became extinct. From approximately 280 Ma ago until 65 Ma ago, dinosaurs and other reptiles 'ruled the Earth'. Few of us remain unimpressed by the images of *Tyrannosaurus*, *Triceratops* and other giant reptiles presented in a vast array of films, books, toys, computer graphics and so on (see Figure 2.10). If we had lived on Earth just before this mass extinction (which of course we didn't – humans didn't arrive on the scene until nearly 65 Ma later), we probably would not have predicted the downfall of these powerful reptilian giants. One prevailing theory is that although biodiversity had been declining for several Ma before this time, 65 Ma ago a meteorite struck the north coast of the Yucatan Peninsula in Mexico, generating a thick cloud of dust that enveloped the Earth. Very little sunlight would have penetrated through this, and the immediate effects could have included a drastic change in climate, tidal waves and extensive fires. The earlier mass extinctions were probably caused by several factors including climate change, reduced oxygen levels and changing sea levels. However, although analysis of rocks has given us some idea about the climatic conditions during these events many millions of years ago, we are far from having the full picture and may never know in all certainty what caused so many species to become extinct within these relatively short time periods.

From the information that we have gleaned from the fossil record, it seems that mass extinction periods were typically followed by a period of between 5 and 10 Ma of very low biological diversity. After this time, when diversity recovered, the plants and animals were often very different from those that were dominant before the mass extinction. For example, when dinosaurs 'ruled the Earth', mammals were relatively scarce and generally kept a fairly low profile. After the dinosaurs had been wiped out, mammals started to flourish, evolving from a small number of types into the diversity of forms including flying mammals (bats) and aquatic mammals (e.g. whales) that exist today.

2.4 Evolution and extinction

Although the fossil record is incomplete, it provides us with enough information to conclude that extinction has followed both steady (background) and sporadic (mass) patterns throughout geological time. The appearance and disappearance of species in the fossil record suggest that up to 99 per cent of species that have ever lived are now extinct. How is it possible then that, despite all the extinctions that have occurred, biodiversity on Earth now seems to be higher than it has ever been in the past? The answer to this lies in the process of evolution.

Although estimates based on the fossil record suggest that a greater number of species live on Earth today than at any other time in Earth's history, this is a very small proportion (estimated at between 1 and 4 per cent) of the species that have ever lived. The total number of species that have existed since life began may be as high as 100 times the number of species living today. The fossil record tells us that these species did not all emerge at once and gradually start going extinct (refer back to Figure 2.6). Over the past 3.5 billion years there has been a steady loss of species due to extinction, and an even greater increase in species due to evolution. Does this mean that we should not worry about extinction because whenever a species becomes extinct, another species evolves to take its place? Before we can answer this question, we must understand a little bit about how species evolve.

How does evolution occur?

It is easy enough for us to understand that if all members of a particular species die, that species becomes extinct. Evolution, the process that has led to the increasing number of species throughout the history of life on Earth, is less straightforward. In order to identify some of the key requirements for evolution, we will consider the example of the peppered moth, *Biston betularia*.

(a)   (b)

Figure 2.11 The peppered moth *Biston betularia*: (a) light and dark forms against a tree with light bark (unpolluted), and (b) light and dark forms against a tree with dark bark (polluted).

The peppered moth in England is found in two forms: a light-coloured form and a dark-coloured form (see Figure 2.11). During the mid-1800s the light form was predominant in the English countryside. By 1898, however, the situation had reversed and the dark form was the most common type of moth in the English countryside. Researchers noted that the spread of the dark form occurred during a time when industrial pollution increased. In order to understand why the dark moths were becoming far more common than the light moths, the researchers constructed a hypothesis. A hypothesis is an idea based on one or more facts or observations that is used as a starting point for an investigation. In this case, the researchers noted that the dark moths were better camouflaged than the light moths when the moths were settled on tree trunks that had been darkened by soot from factories (Figure 2.11). Camouflaged moths should be less visible to birds and therefore the researchers hypothesized that the dark moths on the sooty trees survived better because they were less likely than the light moths to be eaten by birds.

Researchers tested their hypothesis by releasing both dark and light moths in polluted parts of the countryside, and then observing which moths were more likely to survive. In keeping with the hypothesis, the light moths died much more rapidly than the dark moths because they were very conspicuous on the dark trees and were therefore easy targets for the birds that preyed on them. Additional support for this hypothesis came from non-industrial regions (and areas upwind from polluters) where the trees were lighter, because in these regions the light moths were the least conspicuous and therefore survived longer and greatly outnumbered the dark forms.

Evolution by natural selection

The change from light moths to dark moths in many parts of the British countryside was an example of evolution occurring as a result of natural selection. The theory of evolution by natural selection originated in the middle of the nineteenth century with the independent work of natural historians Charles

Darwin and Alfred Russel Wallace, and remains highly influential in the field of evolutionary biology. To try to understand how natural selection works, it will be helpful to move forward to 1952, when an experiment by biochemists Alfred Hershey and Martha Chase confirmed the role of **DNA** (deoxyribonucleic acid)   DNA in inheritance, a finding that provided an explanation for how characteristics could be passed on from one generation to the next. DNA is found in all living things (with the exception of some viruses), and is an essential building-block of life. DNA is passed on from parents to offspring during reproduction, and the arrangement of DNA determines to which species an individual belongs – it is DNA that determines whether an individual is a jellyfish or a tree or other type of organism. In each individual, DNA is arranged into units called genes. Each gene or group of genes is responsible for a different structure or function; for example, in humans, eye colour is determined by a particular group of genes, skin colour by another group of genes, and so on (although note that many traits are influenced by a combination of DNA and environment). In the case of the peppered moth, a single gene determines whether an individual moth will be light or dark in colour. This means that dark moths will pass on the dark colour gene to their offspring, and light moths will pass on the light colour gene to their offspring.

The theory of natural selection as it is known today is based on three major criteria:

- Variation in the characteristics or traits shown by individual organisms within a population. In the above example, this requirement was met because some moths were dark and some were light.
- Higher reproductive success (defined by producing greater numbers of viable offspring) of individuals as a direct result of the trait that is evolving. This was evident in the dark moths in the industrial regions, because they were much more likely to survive than the light moths. As a result of their increased survivorship, they were more likely to mate and produce offspring.
- The traits must be genetically determined and heritable. We know from research that the colour of the moths was controlled by a single gene. This means that adults can have offspring with the same characteristics as themselves – dark moths will have dark offspring – and therefore the trait is heritable.

'Evolution by natural selection' is called a theory, not a hypothesis: a theory is an explanation for something, and is more robust than a hypothesis because it is based on both facts and principles for which some patterns of cause and effect have already been established. The theory of evolution by natural selection is based on a substantial body of evidence, including observations and collections, a great many studies of different species in natural and laboratory conditions, numerous mathematical models, plus a knowledge of genetics.

Natural selection is not the only way in which evolution can occur. For example, neutral changes (changes that are neither harmful nor beneficial) may also occur in an individual. If these are heritable, they too may be randomly handed down from parent to offspring. An important aspect of evolution is that it owes a lot to

DNA is made up of four different types of molecules called nucleic acids, and these are represented as G, A, T and C. DNA is subdivided into genes, and each gene is made up of a specific sequence of nucleic acids that affects the way in which a gene is expressed. If a gene is altered and the sequence of nucleic acids is changed, we say that a mutation has occurred (mutation = change, alteration).

chance, meaning that individuals cannot 'choose' to evolve in a particular way. It was a chance genetic mutation that first created dark moths in the example just discussed, although the improved survival of dark moths in the industrial area was due to natural selection, which is *not* attributed to chance. This initial element of chance means that the speed at which evolution occurs is highly variable (for example, it may take one year or one million years), because it relies on the occurrence of a random mutation that will benefit an individual in a particular environment.

A final point to note about evolution is that it is specific to an individual's surroundings. As we saw in the above example, the dark moths did not do well in the non-polluted parts of the countryside – evolution from light to dark moths was seen predominantly in areas in which there was an advantage in being dark. In order to fully appreciate how evolution may have produced many millions of species over time, we must think about the phenomenal diversity of environments on Earth that have been continually changing since life began. If we imagine millions of species adapting through evolution (or failing to adapt) to such a variety of changing environments over many millions of years, then we can start to understand how so many species have evolved and become extinct throughout history.

Activity 2.3

What conditions must have been met before giraffes could evolve long necks?

You will find the answer to this activity at the end of this chapter.

Can evolution counteract extinction?

If a random genetic mutation allows an individual to produce more offspring, and if that mutation is passed on from parents to their offspring, then over time we would expect all individuals within a population to carry that mutation. This is the way in which evolution occurs, because eventually all individuals within a species will have the characteristic(s) that are determined by the 'new and improved' gene. It is difficult to pinpoint exactly when evolution results in the formation of a new species, particularly if the changes are small and accumulate over a long period of time. One way to think about this is to compare the newly evolved population to the populations that it used to resemble (which, to complicate matters, may no longer exist). If it seems unlikely that individuals from the 'new' and 'old' populations would be able to interbreed, then we would suggest that a new species has evolved. In our example, the dark and light moths are still considered to be in the same species because the genetic differences between the two groups are minimal, and they continue to interbreed.

As we have learned, the speed at which favourable mutations will occur, and therefore the speed at which evolution will proceed, is impossible to predict because it is based on a purely random process. What does this mean for extinction and biodiversity? Well, if every time a species went extinct it was immediately replaced by a newly evolved species, then extinction might not be too much of a cause for concern. However, it seems unlikely that random mutations would provide such an efficient process of 'replacement' species. This scepticism is supported by the 5 to 10 Ma of low biodiversity that apparently followed each mass extinction event found in the fossil record. Evolution may occur quite rapidly, or it may take millions of years. While evolution may counteract extinction in the long run by creating new species, it is highly unlikely to act on the large, rapid scale that would be required if the effects of mass extinctions were to be quickly reversed.

2.5 Are today's extinction rates similar to those of the past?

In order to decide whether or not current extinction rates are similar to extinction rates that have occurred in the past, we have to compare the two. While we have a rough idea about the numbers of species that are now going extinct (at least with respect to the better known taxa, such as birds and mammals), it is extremely difficult to assign a precise number to the background extinction rate that has occurred over the past 3.5 billion years. The average lifespan of species in the fossil record has been estimated at 4 Ma, and from this number we can estimate roughly that, if there were about 10 million species in total, the extinction rate would have been about 2.5 species per year. This translates into 1 out of every 4 million species going extinct each year. However, there is a bias in the fossil record, because a species that lived for 10 Ma would have had many more chances to have been included in the fossil record than a species that lived for only 100,000 years. Therefore, the species that appear in the fossil record will be disproportionately represented by long-lived species, and this will lead us to overestimate the average lifespan. Once again we are reminded of the concept of uncertainty with respect to the fossil record.

If we accept that we have overestimated the average life span of species from the fossil record, we must compensate for this in some way. It is not clear how we could do this, but let us assume that the background extinction rate is 10 times higher than the rate estimated above. If we average extinctions over all different types of plants, animals and other organisms then, using this revised, higher background extinction rate, extinctions of the 4,000 or so mammals living today would be expected to occur at a rate of about 1 every 100 years, and of birds at a rate of 1 every 50 years (Groombridge and Jenkins, 2000). If we wish to be less conservative and assume that the background extinction rate is *100 times* higher than the rate estimated above (that is, 1 out of 40,000 species going extinct each year), we can expect current extinction rates of about 1 mammal every 10 years, and 1 bird every 5 years. If 1 bird species is expected to go extinct every 5 years, this translates into 40 bird species over the past 200 years. The reality is that 103 bird species have gone extinct since 1800, which is nearly 2.5 times the rate of the less conservative estimate.

Activity 2.4

Earlier in the chapter we learned that BirdLife International has predicted that 182 bird species have only a 50 per cent chance of surviving the next 10 years. How does this compare with the average extinction rate throughout time?

The response to this question is at the end of this chapter.

Summary

Our comparison of current and historical extinction rates has shown us that species are going extinct today at a much faster rate than they have done previously (with the exception of past mass extinctions). It therefore appears that we do have cause for concern, as we are losing species much more quickly than generally occurred in the past. In fact, the current extinction rates are so high (and predicted to continue to rise) that many people believe that we are entering a sixth mass extinction. Lord May, when he was president of the Royal Society of London (the Academy of Science in the UK), calculated in 2001 that the current extinction rates of birds and mammals is probably 100 to 1,000 times faster than the average calculated over many millions of years of history. This estimate, plus the fact that humans are placing increasing pressure upon Earth's resources, led May to conclude that 'There is little doubt that we are standing on the breaking tip of the sixth great wave of extinction in the history of life on Earth' (May, 2001). The finality of extinction should remind us of the issue of why environmental questions are so pressing.

3 Do extinction rates vary over space?

We will now move away from using time as an analytical tool for understanding extinction, and consider how the notion of space may provide us with additional insight. By examining regions with differing levels of extinction and biodiversity, we gain another perspective on whether or not we should be worried about the numbers of species that are going extinct.

3.1 Space as an analytical tool

In our analysis of historical extinction rates we have been using extinction rates that were averaged over extremely long periods of time, and this method may disguise some of the fluctuations that occurred over shorter timeframes, such as from one century to the next. For example, if we know that 50 extinctions occurred over 500 years, we may conclude that there were 10 extinctions every 100 years. However, if 50 species went extinct in the first century and no species went extinct for the next 400 years, the average extinction rate over that 500 year period would still be 10 species per century. It is reasonable to assume that background extinction rates were variable over the past several billion years, and we cannot prove that they were seldom or never as high as the rate is today (barring periods of mass extinction). However, what we can do is try to reinforce our argument that extinction rates are currently high by using space as an analytical tool. If we see marked differences in extinction rates in different parts of the world, then we can focus on the areas with high extinction rates and acquire some insight into what is causing these extinctions. This should provide further information to help us to assess whether or not current and imminent extinction rates are a cause for concern.

3.2 Biomes and species distributions

Before we compare data about species living in various regions around the world, we must first consider why a species will inhabit any particular area.

○   Is biodiversity evenly distributed around the globe, with different types of habitat home to similar numbers of species? In answering this question, think of areas that have a high number of species, and those that have relatively few species living in them.

●   As far as we know, some areas apparently have much greater biodiversity than others. Tropical forests are perhaps the most well known species-rich areas. Regions with relatively few species living in them include deserts and the Arctic.

Imagine the landscape of the species-rich versus the species-poor regions suggested above. What is the first difference that you see in your mind's eye? Perhaps the most obvious difference is the presence or absence of vegetation. Tropical forests have an abundance of trees, whereas the vegetation in deserts and in the Arctic is generally sparse and very small. The type of vegetation in an area sets the scene for all other inhabitants. This is because plants and other so-called **primary producers**, such as algae, can **photosynthesize**, which means that they can capture energy from the sun and use this energy to build plant tissue. This plant material then provides food for **herbivores** (animals that eat plant material) and **detritivores** (organisms that eat dead organic matter). Herbivores in turn may provide a food source for **carnivores** (animals, and occasionally plants, that eat animal tissue). This relationship between plants, animals and other organisms is called a **food web** (Figure 2.12), and forms a basis for ecosystems (introduced in Chapter One and see **Morris and Turner, 2003**, for a more detailed discussion).

primary producer

photosynthesis

herbivore

detritivore

carnivore

food web

At this point, it is enough for us to know that without plants there would be no food for animals, because animals lack the ability to photosynthesize. This is an important reason why the presence or absence of plants plays a key role in determining how much biodiversity exists in any given area.

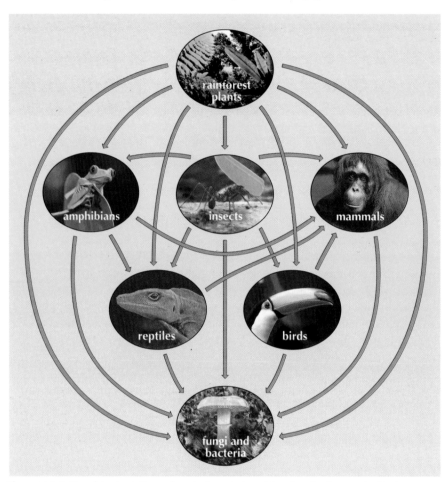

Figure 2.12 Graphic depiction of part of a rainforest food web. Arrows indicate groups that are major consumers of other groups.

What influences the distribution of plants?

biome

The major vegetation types of the world are referred to as **biomes** (Figure 2.13). Each biome is defined by its physical structure, for example whether it contains trees or not, how densely these trees grow together, what other types of vegetation are present, and so on.

Figure 2.14 shows the distribution of the major biomes around the world. It is worth noting at this point that any particular biome can be found in regions of the world that are separated by thousands of kilometres, and that areas of the same biome are not necessarily home to the same set of species. For example, the

(a)   (b)

Figure 2.13 Examples of two different types of biomes: (a) prairie in Montana, USA, and (b) bluebell wood in England.

prairie grasslands of North America and the grassland steppes of central Asia are both representatives of the grassland biome, but each is home to a very different group of species. Figure 2.15 shows that the distribution of biomes around the world is strongly influenced by two factors: temperature and precipitation.

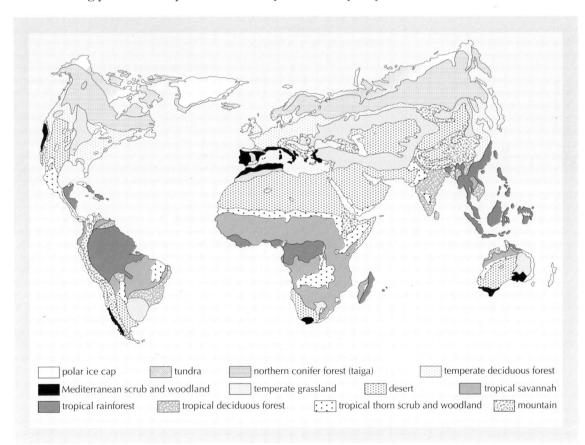

polar ice cap   tundra   northern conifer forest (taiga)   temperate deciduous forest

Mediterranean scrub and woodland   temperate grassland   desert   tropical savannah

tropical rainforest   tropical deciduous forest   tropical thorn scrub and woodland   mountain

Figure 2.14 Distribution of the major types of biomes around the world.

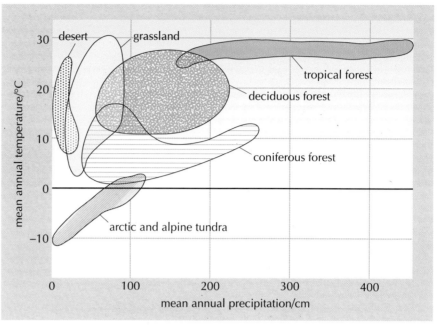

Figure 2.15 Six major biomes related to their mean annual temperature and total precipitation.

○   Think back to the types of habitats found within different biomes, and then use the information provided in Figure 2.15 to determine what combinations of temperature and rainfall produce relatively high and relatively low levels of biodiversity.

●   When both rainfall and temperature are high, biodiversity is high (tropical rainforests). When temperature and rainfall are both low, biodiversity is low (Arctic).

Understanding the distribution of biomes is important if we are to understand how biodiversity is distributed around the world. The tropical rainforest biome contains the highest levels of biodiversity (although this is a generalization – there is extensive variation within and among tropical rainforests). While the exact reasons for this high biodiversity within rainforests are not well understood, it seems that the combination of the temperature and rainfall in these areas is suitable for a wide range of plants that in turn provide food and shelter for a wide range of animals. Tropical forests cover about 7 per cent of the world's surface, and are believed to provide a home for up to 50 per cent of the world's species. Figure 2.16 shows how relative levels of biodiversity are distributed in countries around the world, and Figure 2.17 shows how endangered plant and animal species are distributed around the world.

Activity 2.5

Compare Figures 2.16 and 2.17. Do you see any overlap in the patterns of biodiversity and critically endangered species?

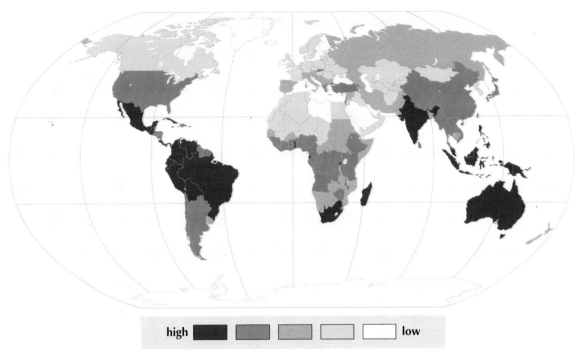

high ▮ ▮ ▮ ▮ ▯ low

Figure 2.16 Relative biodiversity calculated for countries around the world, based on terrestrial (land-based) vertebrates and plants.
Source: Groombridge and Jenkins, 2000.

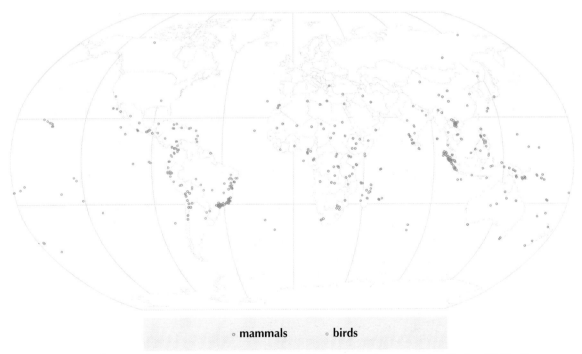

○ mammals   ○ birds

Figure 2.17 Distribution of most of the 337 mammal and bird species that were classified as 'Critically Endangered', the highest risk category, in the 1996 IUCN Red List. Each circle represents one distribution record, so a high density of symbols can represent many records of a single species, or single records of many separate species.
Source: Groombridge and Jenkins, 2000.

Comment

While not in perfect agreement, the distributions of biodiversity and critically endangered species show a fair amount of overlap. There is a general trend for areas of high biodiversity to have increasingly higher numbers of critically endangered species. Note that this figure refers only to *critically* endangered species (those at risk of imminent extinction). The distribution of species for which extinction seems less immediate, but which are nevertheless at risk, is much more extensive throughout the world.

The fact that greater numbers of species appear to be going extinct in areas of high biodiversity may seem unremarkable. After all, if there are more species to start with, then it seems inevitable that there will be higher numbers of extinctions in these areas. Does this mean that we should remain unconcerned about the differences in extinction rates throughout the world? In order to address this question, we will use tropical rainforests as an example.

Why are so many species in tropical rainforests going extinct?

We don't know exactly how many species are going extinct in tropical rainforests, in part because, as we saw earlier, we don't know how many species are there to begin with. According to the World Resources Institute, 100 species become extinct every day due to tropical deforestation. An organization called Save the Rainforest provides a slightly more conservative estimate of 35 species a day going extinct in rainforests around the world. Overall, estimates vary considerably, but the general consensus is that a substantial number of species in tropical rainforests are going extinct every year.

Most species that are currently threatened with extinction are in this situation because of habitat loss or modification, and species found within tropical forests are no exception. Tropical rainforests in many countries around the world are being destroyed at a steady rate, primarily because they are being converted to agriculture and ranching, and are being logged for timber (Figure 2.18). Other threats to rainforests include mining, oil exploration, and the building of hydroelectric dams. The exact rate at which rainforests are presently being destroyed is not known, as there have been no global assessments since 1990. At that time, it was estimated that an area of about 150,000 square kilometres of tropical rainforest, equivalent to the size of England and Wales, was being destroyed every year (Figure 2.19).

Every rainforest in the world is currently under attack. The south-western Amazonian forests of western Brazil, northern Bolivia, and south-eastern Peru, which cover an area of more than 200,000 square miles, present a typical case study. This is a key part of the world's largest intact rainforest – some 94 per cent of the south-western Amazon's original forested area remains forested today.

(a)   (b)

Figure 2.18 Amazon rainforest: (a) before logging, and (b) after logging.

This area, which is considered one of the Earth's most biologically rich, includes lowland tropical moist forests, unique flooded savannas dotted with palm trees, and bamboo-dominated forests covering an area the size of England. The south-western Amazonian forests are one of the last refuges of the endangered jaguar *(Panthera onca)*, the harpy eagle *(Harpia harpyia)* (illustrated in Figure 2.20), and the giant river otter *(Pteronura brasiliensis)*. These forests also have the highest diversity of freshwater fishes, birds and butterflies in the world. As many as 1,200 species of butterflies have been recorded, and many plant species are found nowhere else on Earth. Other inhabitants of these forests are tapirs

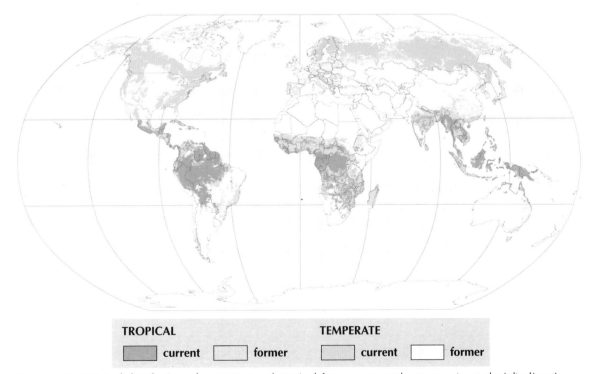

TROPICAL   TEMPERATE

current   former   current   former

Figure 2.19 Original distribution of temperate and tropical forest cover under current (post-glacial) climatic conditions and before significant human impacts, and the distribution of remaining forest. Approximately half of the world's original forest cover has disappeared.
Source: Groombridge and Jenkins, 2000.

(Tapirus terrestris), howler monkeys (*Alouatta* species), pumas (*Puma concolor*), ocelots *(Leopardus pardalis)*, bush dogs *(Speothos vernaticus)* and black caymans *(Melanosuchus niger)*.

(a) (b)

(c) (d)

Figure 2.20 Some endangered species living in the Amazon rainforest: (a) emerald tree boa (*Corallus caninus*), (b) jaguar (*Panthera onca*), (c) harpy eagle (*Harpia harpyja*), and (d) swallowtail butterfly (*Papilio rhodostictus*).

For many years, this unique rainforest had few humans living in it compared to other parts of the Amazon. That has changed in recent decades: human populations in the area are increasing and forests are being converted into farm and pasture. A network of roads constructed to allow access to timber, ore, oil and gas has also made it easier for people to reach remote areas and for fires to travel farther. Satellite images of the 1997 fires in the region show lines of fires that follow roads. Oil and gas reserves are currently being developed in the southern Peruvian Amazon, and there are plans to build a gas line to Bolivia and Brazil. This will bring more development to the region, which means more disturbance and destruction for this environment.

3.3 Habitat loss and species extinction

The relationship between habitat and species is complex. As mentioned in Section 3.2, while species cannot survive without suitable habitat, it is also the case that habitats would not exist without the species that live in them. For example, tropical rainforests would not exist without trees and other plants, which in turn may not exist without birds and insects to pollinate them, detritivores to provide them with nutrients, and so on. As a result, the distinction between habitat and species is somewhat blurred because of the enormous amount of overlap between the two. While we will, largely for the sake of convenience, continue to discuss extinction primarily in terms of species, it is important to remember that we are also referring to extinction of habitat.

Key to any habitat are the species (especially plants) that contribute to its structure, such as the trees that form the forest or the grasses that form the grassland. As these structural species are lost, the habitat is destroyed and many of the animals that coexisted with these plants are also lost. However, there are also other less obvious effects that occur when habitats are altered. For example, even when a forest is not completely cut down, it is often fragmented into small islands, such as those shown in Figure 2.21. This is detrimental to species for a number of reasons.

Figure 2.21 Deforestation has led to fragmented forests in areas of Peru.

First, there is the edge effect. All sorts of species may avoid living at the edge of their habitat for a number of reasons. If we think of the edge of a forest, there may be greater risk to animals of being hunted by species (including humans) that would not normally venture into the forest. There may also be more competition for food from species that would not be found inside the forest. In addition, there may be a change in microclimate (the climate that exists within a small area), as there may be less shade, higher temperatures and more wind when the tree density starts to fall. This may not be too critical for larger animals, but could be extremely important for certain plants and insects. For all of these reasons, the species that live quite happily in the heart of the forest may not be able to survive at the edges. The more fragmented a forest is, the more edges there will be, and the consequences of this could be dire for many species.

The second problem with fragmented habitat is that smaller areas often cannot support as many species as larger areas. This is in part related to the complexity of food webs. For example, we can imagine a herbivore whose diet consists mostly of a plant that is quite common, but that grows at low density (that is, it is spread out over a wide area). This herbivore is in turn the main source of food for a particular carnivore. If the total size of the habitat is reduced, the herbivore may not be able to find enough of its main food plant and will die (as will diverse parasites, bacteria and other micro-organisms that lived in or on this herbivore). In turn, the carnivore will lose its main food source, and it too will die (plus its micro-organisms). The loss of one species will often affect other species at different points in the food web.

A third effect of habitat fragmentation is a reduction in habitat size, which often means that only small populations can be supported. Small populations are generally at greater risk of extinction than large populations.

Why are small populations at greater risk of extinction than large populations?

A population is a collection of organisms belonging to the same species, found within a limited geographical area. The balance of four processes determines whether a population size increases, decreases or remains constant. These processes are birth, death, immigration into the population and emigration from the population. Birth and immigration add new individuals to a population, whereas death and emigration remove individuals from a population. If there is no migration to or from a population, birth and death are the only two factors that determine population size.

Activity 2.6

The following exercise illustrates how a rabbit population may grow rapidly and spread through an area.

Assume that there is a birth rate of 2,500 rabbits per 1,000 adults each year, and a death rate of 700 rabbits per 1,000 adults per year. For the moment, ignore the possibilities of immigration and emigration. If the birth and death rates remain constant, start with a population of 1,000 individuals, and calculate what size this population would be at the end of years 1, 2, and 3.

You will find the answers to this activity at the end of this chapter.

exponential

The above example represents an **exponential** increase in the population size (see Figure 2.22). In the absence of immigration and emigration, we know that when the birth rate exceeds the death rate, there will be an accelerating increase in the population size because each year there will be more adults that will reproduce. If the death rate exceeds the birth rate, the population will become smaller and smaller each year until the population goes extinct.

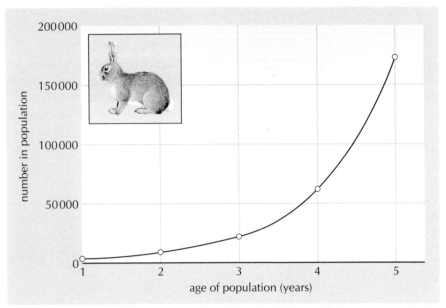

Figure 2.22 Exponential increase in a hypothetical rabbit population over a period of five years.

If the sum of birth and immigration is equal to the sum of death and emigration, the size of a population will remain constant and that population is said to be at equilibrium. However, population sizes seldom remain constant from year to year. For example, in years when food supply is low, mortality may increase and population size will decrease. If a population is small to begin with, then it lacks a 'buffer' that will allow it to survive periods in which the sum of death and emigration are greater than the sum of birth and immigration. This is one reason why small populations are more susceptible to extinction. The other reason relates to the amount of genetic variation within populations, a measurement that was introduced in Section 2 as one way of assessing biodiversity.

The total number of different genes within a population is termed its genetic diversity (Section 2.1 and Box 2.1). Because all individuals are genetically unique (even clones and identical twins, although their genetic differences are minimal), large populations generally contain a large number of different genes. The exact number of genes will of course vary. For example if there is a lot of asexual reproduction then the amount of genetic variation, even in a large population, could be low. In sexually reproducing individuals, the amount of genetic variation will depend to some extent upon whether or not inbreeding occurs. Closely related individuals are genetically more similar to one another than are distantly related individuals. In sexually reproducing species, offspring generally share 50 per cent of their DNA with each parent. Similarly, siblings share approximately 50 per cent of their DNA with one another. For this reason, if two siblings were to produce an offspring (an example of inbreeding), then that offspring would have many identical genes from each parent and would therefore have fewer *different* genes than an individual whose parents weren't related. Small populations often have relatively high levels of inbreeding – there are fewer mates from which to choose and so an individual is more likely to mate with a relative.

Genetic variation within a population can be very important. For example, it is often the case that individuals with certain genes will have resistance to a particular disease that may affect other individuals. If there is an outbreak of this disease, larger populations are more likely to contain individuals with the resistance gene(s), thus increasing the likelihood that larger populations will survive. In a population with low genetic variation there may be no individuals with the resistance gene, in which case a disease outbreak may wipe out the entire population. Therefore, it is not simply the size of a population that should be used to predict its chances of survival, it is also the amount of genetic variation. In some cases, a small, genetically diverse population may be more viable in the long run than a larger inbred population.

Population size and space

As we are concerned in this section with the concept of space as an analytical tool, it seems appropriate that we should now pause to ask how much space a population needs if it is to survive for a long time. This is a very difficult question to answer for a number of reasons. First, there will obviously be great variation between species. For instance, in a rainforest in Belize (a small country in Central America) a beetle population should require much less space than a jaguar population because, unlike jaguars, beetles don't have to hunt large prey items. However, over a prolonged period, the beetle population may require more space than you may at first think.

This leads us to the second reason that space requirements are difficult to quantify – it is impossible to predict what will happen to habitats in the future. Say, for example, that our hypothetical beetle population lived only on a single species of tree. If a fungal infection killed a cluster of that tree species, then the beetle population would have to disperse to a new cluster of trees in order to survive. This introduces a whole new set of variables. How far can these beetles disperse? Do their dispersal abilities vary throughout the year? How common are the trees that they inhabit? Do these trees have specific requirements, perhaps growing only at the edge of the forest?

The beetle example shows that knowing quite a lot about the species in question can help us to estimate how much habitat is required for long-term survival, but this estimate cannot anticipate all possible events. As mentioned in Section 3.3, fragmentation of habitats such as forest, prairies and wetlands is an ongoing problem for many wildlife species, because it often reduces the amount of habitat available to a population. One way in which the effects of habitat fragmentation have been reduced is through wildlife corridors. These provide a strip of suitable habitat joining two fragmented areas, for example a corridor of trees joining two forest fragments, which wildlife can use to travel from one forest area to the other. This is important because it expands the size of the habitat in which a species can live. In Belize there is a network of corridors between fragmented forests, creating a forest habitat large enough for jaguars (Chapter Three discusses wildlife corridors and tiger reserves). Wildlife corridors also provide an avenue of escape if local conditions deteriorate. In addition, they effectively create a single, larger population that should have much more genetic variation than smaller fragmented populations (Figure 2.23), and this improves chances of long-term survival. However, there are also potential drawbacks to wildlife corridors, including an increase in 'edge'.

In the preceding sections we have learned that space is relevant to extinction and biodiversity in a number of ways. At the species level, it is apparent that populations need varying amounts of space, and this is influenced in part by their ability to disperse between sites. Dispersal is, in turn, influenced partly by connectivity among sites, which can be enhanced by relatively simple measures such as wildlife corridors. At the ecosystem level, overall biodiversity is much higher in some areas than in others, and tropical rainforests collectively contain more biodiversity than any other biome. The tropical rainforest habitat is being lost over great areas, largely as a result of human activity, and this loss is leading to the extinction of an unknown (but high) number of species.

Wildlife corridors are just one means of habitat protection promoted by conservation biologists and organizations such as WWF (World Wide Fund for Nature, known as World Wildlife Fund in North America). Groups such as these are working in many areas to try to limit habitat loss, poaching, illegal trade and other factors that contribute to the demise of species.

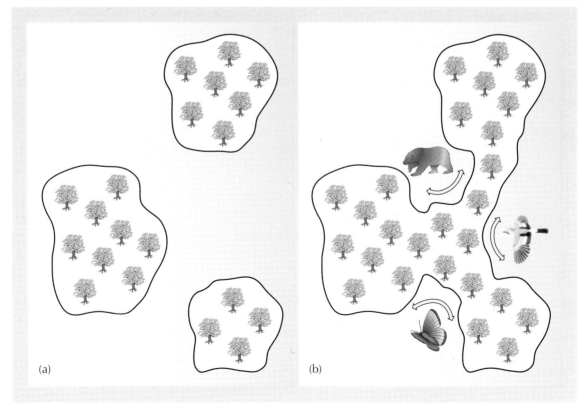

(a)

(b)

Figure 2.23 Schematic diagram showing how suitable corridors can connect previously isolated habitats, thereby increasing the amount of gene flow among locations and enhancing the population size.

Does our example of the rainforest demonstrate that space as an analytical tool has given us greater insight into current levels of extinction? Well, we have seen that, by selectively destroying areas of high biodiversity around the globe, humans are driving large numbers of species to extinction, possibly on a daily basis. However, a counter-argument may claim that extinction will naturally be higher in areas of high biodiversity, simply because there are more species that can go extinct. To consider this we need to examine the relationship between space and extinction in areas of low biodiversity.

3.4 Islands versus mainlands

We will now look at patterns of extinction on two rather different sorts of space: islands and continental masses (mainlands). Many more species go extinct on islands compared to mainlands. In the last 400 years, approximately half of all known mammal extinctions and 90 per cent of all known bird extinctions occurred on islands. Yet the Earth's total land area contains only a fraction of island environments, and they often harbour relatively low levels of biodiversity. Why do so many extinctions occur on islands? In order to answer this question, we will first look at a number of factors that have contributed to some of the key differences between islands and mainlands.

The most obvious difference between a mainland and an island is that islands tend to be a lot smaller. The small size of islands means that there are enough resources (food, water, shelter) for only a limited number of species. In addition, islands are separated from the mainland by expanses of water of varying size, and so they may be geographically isolated. The isolation of islands means that, historically, relatively few species have made it as far as the more remote islands (for example Hawaii, which is 4,000 km from the nearest continent). Therefore, both small size and geographic isolation mean that biodiversity tends to be a lot lower on islands compared to mainlands.

The trend of relatively low levels of biodiversity on islands presents us with what at first glance appears to be something of a paradox: how is it that relatively high numbers of extinctions occur in areas with relatively few species? Part of the reason relates to population sizes and large numbers of **endemic** species (species found in a particular area, such as an island, and nowhere else). Two reasons why so many island species go extinct are:

- Populations are often small, because of the limited sizes of islands (refer back to Section 3.3 on why small populations are vulnerable).
- There are often high levels of endemism (many endemic species) on islands which means that extinction of one population may often mean extinction of an entire species.

While these two factors can explain a number of 'naturally occurring' island extinctions, they do not fully explain why island extinction rates have accelerated in recent years. The answer to this can be found by examining what happens when species that have evolved on islands are exposed to introduced species.

The destruction of tropical rainforests by humans is just one example of how environmental conflict (introduced in Chapter One) may be between humans and non-humans, a situation in which non-humans are most commonly the losers (with the notable exceptions of bacteria and other agents of disease).

endemic

When we say that a species has gone extinct, we don't always mean that it no longer exists anywhere on Earth. It could be locally or regionally extinct, i.e. no longer found in an area that it formerly inhabited. However, this distinction does not apply to endemic species with small ranges, because in those cases local extinction equals global extinction.

Evolution on islands

Species that manage to establish themselves on islands will, over time, continue to evolve. In Section 2.4 we learned that natural selection causes species to evolve in a manner that leaves them well suited to the particular environment in which they are living. A key difference between islands and mainlands is the historical lack of mammalian predators and herbivores on many islands. This absence arose because land mammals can neither swim long distances nor fly (with the exception of bats), and therefore cannot move quickly over the ocean. Furthermore, land mammals cannot survive for long periods without fresh water (unlike some reptiles), and therefore cannot slowly float across the ocean. As a result, mammalian land predators, such as rats or cats, and mammalian herbivores, such as goats, have been absent from many islands.

○   How might the lack of predatory mammals influence the evolution of birds and reptiles on islands?

●   Think of the peppered moth in Section 2.4. If there were no birds eating the moth, then it might not matter whether or not it was camouflaged on the tree bark. Similarly, the absence on islands of predators, such as rats, foxes or cats, may make birds and reptiles less wary about behaviours such as nesting in open areas, because there will be no selection against individuals that are eaten by predators (see Figure 2.24).

Plants, like animals, continue to evolve in response to their environment. One set of plant 'predators' are mammalian herbivores (goats, rabbits, deer, etc.), and plants in some places have evolved a variety of strategies to help them avoid being eaten, including unpleasant taste, indigestibility, toxicity, needles and thorns. However, while plants on islands may be consumed by a variety of insects, they will not be selected to evolve traits that specifically deter mammalian herbivores.

Figure 2.24 A flightless cormorant (*Compschalieus harrisi*) on the Galapagos Archipelago. This species probably lost the ability to fly because it simply did not need to fly: food is obtained by diving and fishing, nesting is done near to the water, and until recently there were no mammalian predators from which it needed to escape. Since land mammals have been introduced onto the Galapagos, the flightless cormorant has dwindled in numbers and is now endangered.

Introduced species

Species that are introduced into a particular region, such as an island, do not naturally occur in that region. The vast majority of so-called introductions, deliberate or otherwise, are the result of human activity. Examples of deliberate introductions include numerous agricultural plants and animals, decorative plants, pets, and animals for fishing or hunting. Unintentional introductions include parasites and pest insects associated with the deliberately introduced species, plus species such as rats and cockroaches that often accompany human settlements. The introduction of species into new habitats presents such a threat to biodiversity that it is generally ranked second in global importance, superceded only by habitat loss. Introduced species are particularly problematic on islands because the species already living there have evolved in isolated settings in which there has been no need for defence against mammalian predators, many types of insect pests and other hazards that have previously been absent from the island.

The impact of introduced species on island biodiversity is illustrated by the situation on the Hawaiian Islands. As a result of its extreme isolation from mainlands, Hawaii has only one native land mammal, the Hawaiian hoary bat. The islands of the Hawaiian Archipelago are 5–70 Ma in age, and are old enough for a multitude of species to have evolved as endemics. However, a significant proportion of Hawaii's biodiversity has been lost over the past 200 years. Nearly 40 per cent of birds unique to Hawaii are extinct and, of the 42 species that remain, 30 are classified as endangered or threatened. Forty per cent of native plants have already been designated as endangered or are candidates for such classification. Approximately 900 species (71 per cent) of about 1,263 historically described species of Hawaiian land snails are extinct. Although Hawaii comprises only 0.2 per cent of the US land area, the state accounts for more than 70 per cent of extinctions in the country and harbours more than 25 per cent of the nation's rare and endangered birds and plants.

Many of the extinctions on Hawaii have been attributed to introduced species including the black rat, *Rattus rattus*, which has probably had the greatest effect on endemic plants and animals by eating a wide variety of things including snails, insects, bird eggs, nestlings, fruits and flowers, as well as stripping branches from bushes and trees. Wild goats (*Capra hircus*) have destroyed entire populations of native plants that, in the absence of large herbivores, had not evolved any

'Feral' refers to species that were once domesticated, but have since escaped and now survive as wild species.

resistance to intensive grazing. Goats, along with feral pigs (*Sus scrofa*), cause extensive damage in endangered habitats such as rainforests and montane bogs by eating plants, churning up ground, hastening soil erosion, and introducing into these areas certain non-indigenous insects, parasites and plant seeds that hitch-hike on the pigs. Each year more and more species are being introduced – by 1995 an estimated 2,600 insect species and 861 plant species had been introduced into Hawaii (Howarth et al., 1995). The vast majority of these introductions can be attributed to human behaviour (see Figure 2.25).

(a)   (b)

Figure 2.25 Hawaii: (a) uncultivated land, compared with (b) area cultivated with sugar cane.

Summary

In Section 3 we have used two examples – rainforests and islands – to look at how patterns of extinction may vary in different parts of the world. The first of these, rainforests, provided us with three key facts:

- According to the knowledge that we have at this point, tropical rainforests collectively contain more biodiversity than any other biome.
- The tropical rainforest habitat is being lost over great areas, and this loss is leading to the extinction of an unknown (but high) number of species.
- The destruction of tropical rainforest is largely a result of human activity.

From this information we can conclude that by selectively destroying areas of high biodiversity around the globe, humans are driving large numbers of species to extinction, possibly on a daily basis.

The second example, islands, illustrates how species living in a relatively small and isolated area are often highly susceptible to environmental change, and therefore may be particularly prone to extinction. The recent acceleration of island extinctions can be attributed primarily to the loss of isolation on islands. This has led to the human-mediated introduction of thousands of species, often at the expense of species that were already living there.

As is the case for the tropical rainforests, we see evidence that high extinction rates on islands owe much to human activity. This further reinforces our conclusion that the current high levels of extinction rates are not 'natural', but instead are human-driven and – given the increasing human population (**Drake and Freeland, 2003**) – are unlikely to slow down in the near future.

4   Conclusion: a sixth mass extinction

At the start of the chapter we saw various facts and figures that seemed to suggest that extinction rates are currently very high. We have used time and space as analytical tools to provide evidence in support of this suggestion of high extinction rates. First, an analysis of extinction rates over time revealed that although there has been a fairly steady background extinction rate over millions of years, there have also been at least five mass extinctions in geological history in which overall levels of biodiversity were drastically reduced. These events were followed many millennia later by an overall increase in biodiversity, implying that they are somewhat cyclical. Current extinction rates appear to be much higher than historical background extinction rates. If we accept, as evidence suggests, that we are currently entering a sixth mass extinction, this will be the first extinction caused by a single species, *Homo sapiens*. If previous patterns are repeated then there may be an increase in biodiversity many millennia from now, although whether or not *H. sapiens* will be around to witness this is impossible for anyone to predict.

The second analytical concept used in this chapter, that of space, supported our conclusion that current levels of extinction, at both the species and habitat levels, are extremely worrying, both for those concerned with biodiversity and for those concerned with the fate of the planet as a whole. We examined two types of space – tropical rainforests and islands. The uneven extinction rates in these areas demonstrated that biodiversity is being lost in an unprecedented way, directly related to the actions of humans. Accelerated extinction of both habitats and species must constitute a pressing environmental concern: global extinction is final, and evolution is unlikely to act rapidly enough to 'replace' the extinct species with newly evolved species. In the next chapter, we will start to explore how values, power and action are additional analytical tools that help us make sense of environmental issues, and contribute to our understanding of why these issues are so pressing.

References

Bingham, N. (2003) 'Food fights: on power, contest and GM' in Bingham, N. et al. (eds).

Bingham, N., Blowers, A.T. and Belshaw, C.D. (eds) (2003) *Contested Environments*, Chichester, John Wiley & Sons/The Open University (Book 3 in this series).

Blackmore, R. and Barratt, R.S. (2003) 'Dynamic atmosphere: changing climate and air quality' in Morris, R.M. et al. (eds).

Blowers, A.T. and Hinchliffe, S.J. (eds) (2003) *Environmental Responses*, Chichester, John Wiley & Sons/The Open University (Book 4 in this series).

Drake, M. and Freeland, J.R. (2003) 'Population change and environmental change' in Morris, R.M. et al. (eds).

Groombridge, B. and Jenkins, M.D. (2000) *Global Biodiversity: Earths' Living Resources in the 21st Century*, Cambridge, World Conservation Press.

Hinchliffe, S.J. and Blowers, A.T. (2003) 'Environmental responses: radioactive wastes and uncertainty' in Blowers, A.T. and Hinchliffe, S.J. (eds).

Howarth, F.G., Nishida, G. and Asquith, A. (1995) 'Insects of Hawaii' in LaRoe, E.T., Farris, G.S., Puckett, C.E., Doran, P.D. and Mac, M.J. (eds) *Our Living Resources*, Report to the Nation on the Distribution, Abundance, and Health of US Plants, Animals, and Ecosystems, Washington, DC, US Department of the Interior, National Biological Service.

May, Lord R.M. (2001) 'Scientist warns of sixth great extinction of wildlife' (address to the Natural History Museum), reported in *Guardian Unlimited*, 29 November, www.guardian.co.uk/uk-news/story/0,3604,608510,00.html

Morris, R.M., Freeland, J.R., Hinchliffe, S.J. and Smith, S.G. (eds) (2003) *Changing Environments*, Chichester, John Wiley & Sons/The Open University (Book 2 in this series).

Morris, R.M. and Turner, C. (2003) 'Dynamic Earth: processes of change' in Morris, R.M. et al. (eds).

Raup, D.M. (1994) 'The role of extinction in evolution', *Proceedings of the National Academy of Sciences*, USA, no.91, pp.6758–63.

Wheeler, Q.D. (1990) 'Species-scape', *Entomological Society of America*, no.83, pp.1031–47.

Answers to activities

Activity 2.1

(a) Fishes, amphibians, reptiles and mammals had all evolved by this time, about 206 Ma ago.

(b) (i) In the Devonian
(ii) In the Permian

(c) In the late Jurassic

(d) This figure tells us that new species continued to emerge throughout the Phanerozoic, with an apparent overall increase in biodiversity.

Activity 2.3

There must have been:

(a) A heritable mutation for longer necks

(b) An advantage to long necks within the giraffe's environment (for example, the ability to access a previously unobtainable food supply such as leaves on higher branches)

(c) Increased survival and reproduction as a result of this advantage

If all these conditions were met, natural selection for long necks would have occurred and longer necks would have evolved.

Activity 2.4

If 182 bird species have a 50 per cent chance of survival over a 10 year period, we may expect half of these to survive, in which case 91 species will go extinct. If 91 species go extinct over 10 years, this rate would mean that approximately 45 will go extinct over 5 years (if the rate remained constant). This is much higher than estimates presented in the text, that were based on historic background extinction rates (these estimates concluded either 1 species every 50 years or 1 species every 5 years).

Activity 2.6

For every 1,000 rabbits, add 2,500 rabbits per year for the birth rate, and subtract 700 rabbits per year for the death rate.

At the start of year 1, there were 1,000 rabbits and therefore by the end of year one there had been an increase in 2,500 rabbits and a decrease in 700 rabbits for a total of 1,000+2,500-700 = 2,800.

At the start of year 2, there were 2,800 rabbits. 2,800/1,000 = 2.8, therefore by the end of year 2 there had been an increase of 2.8(2,500) = 7,000, and a decrease of 2.8(700) = 1,960. 2,800+7,000-1,960 = 7,840 at the end of year 2.

At the start of year 3, there were 7,840 rabbits. 7,840/1,000 = 7.84, therefore by the end of year 3 there had been an increase of 7.84(2,500) = 19,600, and a decrease of 7.84(700) = 5,488. 7,840+19,600-5,488 = 21,952 rabbits at the end of year 3.

Chapter Three

Who cares? Values, power and action in environmental contests

Steve Hinchliffe and Chris Belshaw

Contents

1   Introduction: how the passenger pigeon was killed off

Imagine this: a threatening cloud on the horizon, a rumble and then, as the sky darkens, instead of the expected storm, a flock of birds passes overhead. In the USA, passenger pigeon (*Ectopistes migratorius*) flocks were often mistaken for tornadoes. The ornithologist Alexander Wilson witnessed one flock 240 miles long. Writing in the early 1800s, he estimated that there were two thousand million birds in the flock (Price, 1999). Similar accounts of passenger pigeon populations migrating between feeding grounds are well chronicled right up to the middle of the nineteenth century. While the actual numbers are difficult to verify, these accounts may well be about right. It may come as a surprise, then, that by 1880 there were only a few thousand passenger pigeons left. The last one, Martha (see Figure 3.1), died in captivity in 1914.

Figure 3.1 Martha, the last passenger pigeon (*Ectopistes migratoris*), died, aged 29, at Cincinnati Zoo in 1914. The population of this species went from billions to extinction in less than 100 years.

In order to understand why this kind of environmental change occurs we need to adopt an interdisciplinary approach. As the last chapter demonstrated, an understanding of population and habitat dynamics is vital – and this knowledge might even have helped to save the passenger pigeon. Chapter Two also highlights that human actions are at least partly responsible for many extinctions, as is the case with the extinction of passenger pigeons. It is important, therefore, to understand what people can do to their environment and what motivates them to act in particular ways. What people, and perhaps other species, do (their actions) depends to a large extent on what is valuable to them (their values) and what they are capable of doing (their power). This chapter will both explore the interaction between values, power and action, and discuss the importance of these key concepts in relation to environmental issues.

We can get a feel for the importance of these three key concepts very briefly here. It seems clear, for example, that the extinction of the passenger pigeon was brought about in part because the birds were valued more for their meat than for anything else. Even though people like Wilson marvelled at the size of passenger pigeon flocks, many more people valued pigeons instrumentally as a source of food (see Chapter One, in which you were introduced to the idea of instrumental values). Perhaps if more people had valued the spectacle of a large flock as something worthy of preservation, the species would still be with us. How we value environments and components of environments greatly affects how we treat them. A major aim of this chapter, therefore, is to provide an understanding of how environments are valued.

People's values become more significant if there is the power to do something about them. You may already be used to the idea that people do indeed possess the power to produce noticeable environmental effects. While all species change their environments in order to live and reproduce, human beings are remarkable for the extent to which they can alter environments. This is not to say that humans now completely control their environments, nor does it mean that all humans have the same powers. Just as some people have managed to exert more power over their environments, others have seen their ability to exercise power diminished. The power to alter environments is therefore unevenly distributed, not only between humans and other components of environments, but also among people. In addition we should consider the power relations that enable some to exert power over others, who are in turn sometimes able to resist. Another major aim of this chapter is to understand the operation and uneven nature of power.

The final concept we will explore is action. Two kinds of action are important for this chapter. First there are those actions that are part of day-to-day life and that are seldom given a second thought. They include what people do and just as importantly what people don't do. Killing and eating large numbers of passenger pigeons are examples of this kind of action. As we will see, they are closely related to the kinds of values and powers that exist in a particular society at a particular time. Second, there are those actions that involve an explicit attempt to alter the routine ways of doing things. Action in this sense may involve learning more about an issue, forming interest groups and making use of the conventional political process (voting on issues, lobbying representatives), or bypassing the conventional channels and taking what has become known as direct action. Everyday actions and actions that are taken to change the status quo are both important if we are to understand how environmental issues are generated and how things might change. Understanding action is therefore a final major aim of this chapter.

Our overall aim is to demonstrate the importance and usefulness of values, power and action in making sense of environmental issues. We have chosen to discuss all three concepts because they are strongly interrelated. In the first and last sections of the chapter we will be looking at two major case studies of species extinction and endangerment as a means to illustrate these interrelations. In between we will explore in more detail what we mean when we talk about values and ask what kinds of value might be necessary if we wanted to reduce the rate of extinctions you learned about in the last chapter.

In addition to this broad aim, we will answer the following questions:

1   How have some groups of people become effective in their exploitation of environments?

2   How can we distinguish among values and how do these values affect environments?

3   How does power work to enable some and constrain others?

By answering these questions and gaining understanding of the interrelation-ships between values, power and action, you will be able to approach a wide range of environmental issues with some important analytical tools to hand.

2   Pigeons for profit: changing relations between people and passenger pigeons

The causes of the passenger pigeon extinction are not fully known, although it is highly probable that their demise was the combined result of over-hunting (see Figure 3.2), habitat changes and disease. Nevertheless, environmental historians argue that hunting provides a key to understanding the sudden fall in population (Ponting, 1991; Price, 1999).

Figure 3.2 Shooting pigeons in Iowa published in *Frank Leslie's Illustrated Newspaper* in 1867.

The concept of a **resource** is further examined in **Reddish** (2003).

Hunting is a form of **resource** harvesting or exploitation. The term 'resource' refers to substances, organisms and properties of environments that have been defined as useful or valuable for an individual or group. What counts as a resource may change over time with changes in technology and population. While the term 'human exploitation' is sometimes used in a negative sense, all organisms exploit and possibly change their environments in order to survive. They do so by identifying and harnessing the components of their environments that are useful to them. Hunting is an ancient activity for humans and one that is common to many animal and even a few plant species. In this section we will describe how the human harvesting of the passenger pigeon resource changed over time, from an intermittent folk activity to one that was arranged along industrial lines. In doing so, we will note how these changes were accompanied by shifts in the value of passenger pigeons and in the relative power of humans in relation to their environments.

Early hunting

In the early nineteenth century, when passenger pigeons were still numerous, the wonderment that those flocks inspired in ornithologists was matched by the excitement of a pigeon hunt. The appearance of a pigeon flock over a settlement signalled a time for great revelry and a rich harvest of passenger pigeon meat. The flocks would arrive roughly once every seven years, taking advantage of bumper crops of beechnut and acorn. These occasions were known as 'pigeon times', a period of several days when pigeons were caught and festivities were held. 'When pigeons flew over settlements, people followed. Depending on the season and local custom, American colonists shot pigeons, netted them, knocked them down with sticks, ignited pots of sulphur beneath them, or chopped down trees stacked with eighty or a hundred nests apiece' (Price, 1999, pp.8–9).

As the environmental historian Jennifer Price (1999) goes on to argue, these pigeon times were not new. Many indigenous societies had hunted pigeons in large numbers for centuries. One cultural group, the Seneca, an Iroquoian tribe who inhabited the area that today is known as western New York state and Pennsylvania (see Figures 3.3 and 3.4), had been hunting pigeons for at least 700 years prior to the colonists' arrival.

Like the colonists, the Seneca treated passenger pigeons as a food resource for human use, and would kill large numbers. In some ways their hunting was more intense. Rather than waiting for the passenger pigeons to appear every seven or so years, the Seneca would travel large distances to track the pigeon flocks every spring, using this annual migration to meet up with other clans for festivities and for discussing tribal business (Price, 1999). The hunts and meets could last for many days. But this was not an indiscriminate kill. Only young pigeons were allowed to be taken, a rule that effectively, if not necessarily intentionally, maintained the breeding population.

Activity 3.1

It is useful before reading on to summarize the passenger pigeon hunting by both the colonists and the Seneca under a number of headings. This will help us in a more complex activity later in the chapter.

- How often and for how long did the hunts take place (temporal dimensions)?
- Where did they occur (spatial dimensions)?
- What kinds of things did people enjoy about passenger pigeons?

Comment

Our answers would be that the hunts were seasonal and annual for the Seneca, and slightly less regular and more infrequent for the colonists. In both cases the

Figure 3.3 Seneca tribal Chief, 1796.

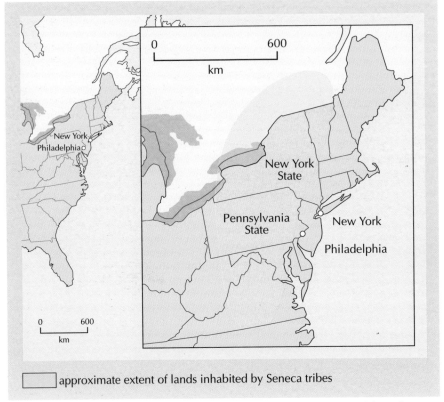

approximate extent of lands inhabited by Seneca tribes

Figure 3.4 Seneca tribes – with modern-day state boundaries indicated for reference.

hunts were carried out over several days. The hunts tended to be local for the settlers while the Seneca travelled relatively large distances. In both cases the pigeons were a valuable source of food, but they were also part of local custom and ritual and a good excuse for festivities.

Hunting and trading: cities and railroads

Pigeon hunts started to be organized rather differently as the nineteenth-century progressed. As more and more people moved to cities, and as capital and labour started to be organized along industrial lines, so the times and the spaces of the pigeon hunt altered.

By 'industrial lines' we are referring to an increase in the scale of organization. Also, the people involved in the industry had specialist jobs in passenger pigeon hunting and processing. People no longer caught pigeons for their own needs, but to exchange or sell in large numbers to make profits.

As live pigeon flocks started to disappear from the immediate vicinities of cities, and as the railways helped to open up new frontier lands (see Figures 3.5 and 3.6), the pigeon trade expanded westwards.

Figure 3.5 North Pacific railroad construction in 1875: railroad construction in the USA opened up the 'Wild West', allowing people and goods to be moved quickly and safely over large distances.

Pigeon carcasses were transported by rail from Milwaukee to Chicago, and on to Philadelphia and New York (see Figure 3.6). Dollars flowed back in the opposite direction. By 1850, several thousand people were employed in the passenger pigeon industry. In New York, in 1855, one enterprise processed 18,000 pigeons each day. A billion birds were harvested in one year in the state of Michigan alone. In the 1870s pigeons were selling at fifty cents a dozen in Milwaukee and two dollars for the same clutch in New York (Price, 1999, p.19). This price differential was the potential profit that could have been made by moving pigeons across the continent. Pigeons were big business and flocks had become valuable economic resources.

Activity 3.2

Using the contour maps in Figure 3.6, calculate how long it took to travel from Chicago to New York in 1830 and 1857. How do you think this increase in speed would have affected the economics of the passenger pigeon industry?

Comment

The journey time shortened from between two to three weeks in 1830, down to one to two days in 1857. The answer to the second question is more open, but long journey times would make the transport of pigeons almost out of the question (especially in summer). Short journey times would mean that the carcasses arrived relatively fresh and would command a higher market price.

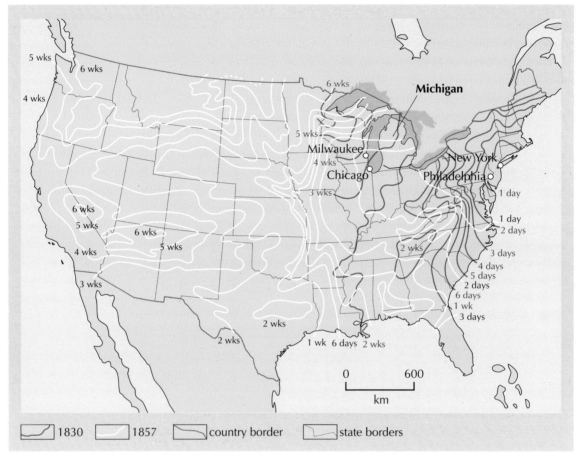

Figure 3.6 Journey times to and from New York, 1830 and 1857.
Source: Cronon, 1991.

An important driving force of this increasingly efficient killing machine was the growing human population in New York and other eastern cities. Passenger pigeons were a favoured form of meat, and even when their availability started to decline, the demand for the sweet-tasting bird stayed high. Indeed, eating pigeons became a mark of high culture and class. Just as importantly, for the wealthy diners in New York restaurants, who were largely unaffected by price rises, pigeon scarcity was not a daily worry. New Yorkers could dine without giving much thought to the problems of finding flocks at the other end of the railroad. The environmental changes that they were contributing towards were very much out of sight and out of mind.

Activity 3.3

In Activity 3.1 you answered a set of questions on the time and space elements of the European colonist and Seneca passenger pigeon hunts. A similar set of questions can now be posed on the passenger pigeon industry.

- How frequent were the hunts that formed part of the passenger pigeon industry?

- How would you characterize the spatial extent of the passenger pigeon industry?
- Passenger pigeons had previously been valued as both food and sport, how did the development of the new industry affect the way that the pigeons were now valued?

Comment

Pigeons were hunted throughout the year, and thousands of pigeons were processed each day. The pigeon trade spanned half a continent. Meanwhile, the pigeon industry viewed pigeon flocks as earnings. Each pigeon had a potential money value.

We can therefore talk about passenger pigeons as **commodities** (as objects produced for exchange rather than personal consumption; in a market economy, this normally means that they are exchanged for cash). We can summarize some of these changing relations between people and passenger pigeons in Table 3.1.

commodities

Table 3.1   Changing relations to passenger pigeons

	Seneca and early Colonists	Passenger Pigeon Production Industry	New York consumers
Pigeons seen as	a food resource Time for festivities	dollars/ commodities	a sign of high society
Temporal and spatial organization	seasonal hunt, regional or local in extent	all year-round industry, near-continental in extent	available all year round, but possible to consume without concern as to origins

Over the course of less than a century, on the land occupied by the booming cities of north-east USA, where once there had been periodic visits from vast passenger pigeon flocks, there was now a regular flow of pigeon carcasses.

2.1 Values, power and action and the passenger pigeon

It will be instructive to consider what this case starts to suggest in terms of values, power and action.

We will have a good deal more to say about values in the next section so, for the time being it is sufficient to note that passenger pigeons were valued in a number of ways. They were valued as an awe-inspiring sight, a food resource, as part of festivities and as commodities for exchange.

Activity 3.4

Refer back to Chapter One where you came across instrumental and non-instrumental ways of valuing aspects of environments. Of the four reasons for valuing passenger pigeons listed in the previous paragraph, which are more like instrumental values and which more like non-instrumental values?

Comment

When passenger pigeons are valued simply for food or as a means of raising money, then they are being valued instrumentally. When they are being valued as a spectacle they are being valued non-instrumentally. Their role in festivities might be more difficult to determine, but this is probably an instance where instrumental and non-instrumental values are mixed together.

It is possible for all of these values to co-exist at the same time, and even for one person to be inspired by passenger pigeons and then to set about earning money by hunting as many as possible. The question of whichever value predominates has important implications for the fate of the species. As a foodstuff (and therefore having instrumental value) the birds would have been worth more dead than alive; and as a spectacle (and therefore non-instrumentally speaking) they would have been worth more alive than dead. In this sense, values affect actions, which then have environmental results. To understand why some values can have a greater impact than others it is useful to understand how power works.

In Chapter One you learned that power was important when it came to various groups seeking to promote their interests. It also became especially important when there were conflicts over interests. You learned how groups might have access to sources of power, including economic resources, political leverage, influence over media and so on. You also learned that a group's power could vary over time and between places, depending on prevailing values and on economic circumstances. So, in any analysis of power we would need to look at the resources that made someone or something powerful, and the context (or what we would call discourses) within which they were trying to further their aims. Power is a complex topic but here we want to introduce some simple means for understanding how power works in a variety of situations.

Power resources

To analyse how power works we need to examine the build up and linking together of various resources. Resources can include people, animals, plants, technologies and skills as well as the political and economic resources that were referred to in Chapter One.

In the passenger pigeon industry's case, we could talk about its power to change environments. The resources included nets and guns, the railroads, the factories, the large workforce, the money markets, the consumers – all of which combined to produce a speedy and efficient process for killing, moving, processing and selling passenger pigeons. You can get a sense of how effective this build up was by comparing the industry's rate of killing (the result of all the resources that it mustered) with the more piecemeal affair that characterized the early colonists' hunting. Another way of thinking about this is to consider how difficult it would have been to stop the industry, with all those business interests tied to the pigeon market. To stop the pigeon trade would have possibly incurred the wrath of the railroad owners, the factory workers, the restaurants owners and the hunters – all of whom had their livelihoods tied to the pigeon industry. This is a typical account of how power works. There is a build-up of resources (allies and organizations) and, if deployed successfully, this power assemblage becomes difficult to counter.

Figure 3.7 The consumption of environmental goods can still proceed without giving much thought to the consequences.
Source: Eales, 1991,

Power discourses

Above, we noted that power is rarely an absolute issue; it is dependent upon the place and time, the values, and not least the relative power of other groups. In short, even significant amounts of industrial, economic and even military power resources may add up to very little if the context is not right. Conversely, those who have mobilized resources may well find their power is enhanced if the prevailing conditions are favourable.

In the passenger pigeon case, the indiscriminate killing was enhanced by the general belief that the bird was so abundant it could not possibly be hunted out of existence. The resource was regarded by many as limitless. Indeed, rumours of the bird's demise were greeted with bemusement, and many expected more passenger pigeons would be found over the next horizon (Price, 1999). Another important consideration that prolonged the large-scale hunting was that people could consume resources like the passenger pigeon without giving much thought to the ways in which the birds reached their table (a style of consumption humorously portrayed in Figure 3.7). The killing was out of sight and out of mind.

One way to capture the importance of such contextual information is to use the term **discourse**. This is a term used to describe practices that inform the way people make sense of the world around them. The practices can include the way in which people talk about issues as well as the everyday routines that colour their understandings and values. Once established, discourses can enable people to make sense to one another, but can also limit the extent to which consideration is given to new or alternative ways of doing things.

You can learn more about the operation of **discourse** in Bingham (2003), and in Maples (2003).

Activity 3.5

Imagine you were keen to save the passenger pigeon in the late nineteenth century. What powers would you need to overcome in order to further your cause?

Comment

You would have to take on the assembled resources of the industry and possibly try to find ways of satisfying the industry's interests by other means. You would also need to consider the habits, norms and values of the society at large, with its strong belief in, or discourse of, unlimited natural resources, and its tendency to consume without needing to worry about origins or production.

It is important to analyse discourses as well as power resources if we are to grasp why certain actions are more likely to succeed than others and how it might be possible to change actions in the future. We can summarize this dual approach to power in the following table.

Table 3.2   How power works

	Resources	Discourses
Power typically works through:	the ability of interests to mobilize and deploy resources (technologies, capital and people) to achieve their objectives	daily practices, norms and habits, the establishment of ideas over what is and what is not possible
In the pigeon example this translates to:	hunters, factories, railroads, wages, workers, wholesalers, retailers, consumers working together	the spatial separation of consumers from 'pigeon production', the idea of a limitless resource

2.2 The end of passenger pigeons

By 1880 the commercial hunting of passenger pigeons had become unprofitable. Flocks were too hard to locate, too widely dispersed and the numbers of birds were too small to justify the transportation costs. Several thousand birds did survive until after this date and it might be expected that this was a large enough breeding population for the species to be able to recover; indeed many species can recover from a comparatively small stock. However, as you read in the last chapter, small populations are often at greater risk from habitat changes, disease and inbreeding. In this case, the passenger pigeon population may already have reached the point of no return. In short, the small, geographically dispersed populations were not sufficient, or sufficiently proximate, to be able to interact to the required degree with one another. It is possible that the dispersal of flocks made the formation of large migratory flocks unlikely. These large flocks may have been vital to the reproduction of the pigeon. Perhaps large flocks enhanced the bird's ability to compete with other birds for nesting opportunities? Or the reduction in population intermixing may have radically altered breeding patterns and mating behaviour? In the end the species did not recover and the death rate exceeded birth rate until the species was extinct in the wild by the turn of the century.

At the turn of the century, more and more people started to talk about pigeons and pigeon times in nostalgic terms. John Muir, one of the first to argue for the preservation of the American wilderness, stated in 1913 that, 'No other bird had seemed to us so wonderful' (quoted in Price, 1999, p.5). This past tense, and sense of loss, fed into a growing realization that the American continent was finite – and the dynamics of wildlife, nature and environment were all far more uncertain than had been assumed. The extinction of the passenger pigeon became emblematic of what many saw as the decline of American innocence. Importantly, and as we will discuss in more depth in the following section, passenger pigeons were no longer being discussed as a potential dinner, or even as potential dollars. Other desires and values were starting to come to the fore. However, although more people were beginning to believe that species like the passenger pigeon had non-instrumental value and were worth preserving, it was not necessarily possible to act on those changed values. The reasons for this are complex, but may have something to do with the build-up and uneven distribution of power. We will return to this in Section 3. You should now have some answers to the first of our questions: how have some groups of people become effective in their exploitation of environments? By now you might also have answers to our second question: how can we distinguish among values and how do these values affect environments? In the next section we build on your current understanding of value to enable you to expand on your answer.

Summary

The following activity is designed to help you check your understanding of Section 2. The activity will prompt you to reflect upon how some people have become effective in their exploitation of environments.

Activity 3.6

(a) What temporal and spatial changes in the organization of passenger pigeon hunts, from early colonial times to the middle of the nineteenth century, allowed North Americans to become highly effective exploiters of their environments?

(b) Did the passenger pigeon industry value the passenger pigeon instrumentally or non-instrumentally? What effects did this valuation have on the bird's population?

(c) How do the notions of power resources and discourses explain why the passenger pigeon went extinct so quickly?

Comment

Indicative answers to these questions are:

(a) The hunting of passenger pigeons changed from being a periodic, local activity, to a year round, spatially extensive activity. Increasingly, the hunting of passenger pigeons took place out of sight of the growing population of urban consumers.

(b) The passenger pigeon industry valued the birds instrumentally as meat and therefore as a means to raise profits. The bird became a commodity and this resulted in a rapid and indiscriminate exploitation of the resource and a substantial decline in the species' population.

(c) The industry was made up of powerful resources including technology, money and people. This allowed for a rapid and profitable harvesting of the species' population. On top of this, a discourse of unlimited natural resources and an ability to consume without giving thought to the consequences, meant that few considered the effects of the rate of exploitation.

3 Values and environments

We have started to open up some of the ways in which values, power and action are involved in and help us to understand environmental issues. As we have seen, the ways in which we value environments can have important effects on the kinds of actions that are taken in those environments. We have used a distinction, made in Chapter One, between instrumental and non-instrumental values, and have shown that what became the most powerful group in the passenger pigeon story valued the species as meat and as a means of making money. For this and other reasons, the species' days were numbered. In this section we will revisit and build on the distinction between instrumental and non-instrumental values, and introduce other ways of thinking about value. The broad aim is to express more clearly what we mean by value, and in so doing understand more of its importance to environmental issues. In addition, we will be asking the question: what kinds of values are needed if we want to avoid future extinctions?

"He appears to have drastically reduced his carbon dioxide emissions"

Figure 3.8 Some argue that the world has value independently of human beings – and may indeed be a better place without them.
Source: Eales, 1991.

We can start by saying some basic things about value:

- All species have requirements or needs in order to stay alive, reproduce and perhaps enjoy a quality of life. As such, we can talk about an owl valuing its prey, or a chimpanzee valuing the space needed to play and sleep. We could also note that water is valuable to a whole range of living and non-living aspects of environments. Value may not always therefore be confined to the needs or desires of humans, and for something to be valuable there may not be the requirement for a sentient creature to place value upon it (a view mawkishly depicted in Figure 3.8).

- Where humans are concerned we can talk about values when there are things that people want, desire, or think good, either for ourselves, for other people, or for the world.

- Values can change over time and space – earlier it was noted that, as the passenger pigeon went extinct, so more people started to express regret at its loss. Likewise, many people who watch wildlife documentaries in northern Europe value animals like tigers for their dramatic beauty. Others living close to wild tigers in parts of Asia may see them as a threat to livestock and people or, if killed, as a source of revenue.

- Different values coexist – people living in the same place and at the same time can often hold different environmental and other values. While some care deeply about the future of the planet, others from similar backgrounds can be uninterested in its fate. Also, any one person's priorities and desires can change over the course of a lifetime and may even vary from day to day. Finally, and as you saw in Section 2 of this chapter, even when people agree that something is valuable they might disagree about why it is valuable.

- Studying value not only contributes to understanding why things like extinctions happen or why the construction of a new airport runway is allowed to go ahead (as in Figure 3.9), it is also an opportunity to consider personal wishes, ideals or interests. For example, while you may value tigers as wild animals, how would you react if you were faced with a choice between a new hospital and better conservation programmes for tigers? To answer a question like this you might not only need to question how you value tigers, but also the extent to which those values matter in the light of possible conflicts of interest. The chances are that at least a part of your reason for

Figure 3.9 These Manchester Airport environmental protesters agree that a new runway shouldn't be built. But do they agree about why it shouldn't be built? Some may be there to protect their homes and wouldn't mind if the runway was built elsewhere; others may be there as a matter of principle and oppose all developments of this kind.

reading these chapters is that you are worried about environmental issues, uncertain as to how serious the changes are, and concerned about what governments, politicians and we as individuals should be doing about the raft of environmental problems that we have to face. The chances are that you're already asking yourself questions about value, almost on a daily basis, for unless we know what to value we don't know what to do. And unless we know what to do, we don't know how to live.

3.1 How can we distinguish among values?

We can begin by considering a pair of distinctions that are often made in thinking about value: instrumental/non-instrumental and subjective/objective. The first of these pairs will now be familiar. Our discussion here is quite general, and not focused on environmental issues in particular. That's because it's important to see how many of our environmental values have significant parallels elsewhere.

Instrumental and non-instrumental

As we saw in Chapter One and above, when something has value as a means to satisfying some other desire or need then it is instrumentally valuable. It is valued in so far as it helps us to satisfy some other requirement or preference. Someone might, for example, value a car because it allows them to get from A to B. That is, they value the car for its ability to satisfy another desire: to move between work and home. As you may know, cars are rarely valued simply as a means of travel. Some people also value them for the supposed status that they confer on the owner. Again, it can be argued that this is an instrumental form of value. The car is a means to some form of social standing. But a car and other things may well be valuable in a different way. Some people buy cars that they admire for their style and feel. There may be nothing that is instrumental in this kind of value. The thing is therefore valued as an end: it is non-instrumentally valued.

This distinction might already be clear. But here's a test for distinguishing between instrumental and non-instrumental value. Call it the no-substitute test. If something is instrumentally valuable, its value will disappear if there's a (preferable) substitute route to the same end. If we could take a pill for perfect teeth, dentists would be out of a job. Slide rules lost their value with the invention of pocket calculators, and can openers are going the same way, as a result of the increased use of ring-pull tops. But when things are valued for themselves, as ends, they are not substitutable in this way. We want that particular thing, and not some approximation.

Activity 3.7

Here's a list of things. Some of them are usually valued instrumentally, or as a means. Others are usually valued non-instrumentally, or as an end. Which are which? Notice the term 'usually': remember, there are many complex cases. At least two of the cases here are complex.

- a kettle
- skiing
- forests
- tigers
- umbrellas
- chocolate

Comment

First the easy cases. Most people value, or want kettles and umbrellas for instrumental reasons. If you only drink beer, and never go out in the rain anyway, you're unlikely to value such things. Skiing and chocolate, in contrast, are usually valued as ends. People mostly ski for fun, not as a way to get to work. And we mostly eat chocolate because it tastes good, not because we're hungry. The environmental cases here are more complex. Some value forests for timber, for the home they offer to wildlife, or for their role in the regulation of global climate. Others are more concerned with their magnificence, or the particular features of trees that can't be found elsewhere. It is a similar story with tigers. Some value tigers for non-instrumental reasons, believing simply that they just look good (Figure 3.10), while others value them for instrumental reasons: thinking that parts of them have quasi-magical and medicinal uses, or wanting their skins for rugs, or valuing tigers for their role within the complex ecosystems in which they are found.

Figure 3.10 The Indian tiger, *Panthera tigris tigris*. How are tigers valued?

It is important to understand that we should not rush to adjudicate once and for all on whether something is either instrumentally valuable or non-instrumentally valuable. Instead we should note that, for many things, both cases are possible.

Figure 3.11 Tigers are valued in numerous ways, but you might want to think about how they *should* be valued?
Source: Hemley and Miles, 1991.

And, significantly, if one side of the debate starts to get the upper hand then there are consequences. For example, if a forest is regarded as instrumentally valuable, say for its timber, then it may well be logged and disappear, or it may be managed in a particular way that maximizes timber production and little else. However, if the forest's non-instrumental value is taken seriously, say as a rare habitat type, or as a place that is cherished by those who visit it, then its existence is likely to be maintained.

Similarly, we can consider the case of wild tigers. A significant threat to the population of wild tigers is the trade in tiger parts, which are used in Traditional Chinese Medicine (TCM). TCM is an immensely popular form of health care relied upon by many millions of people living mainly in Asia and increasingly elsewhere. Demand for tiger products (see Figure 3.11) may therefore be unlikely to decrease markedly in the near future. One solution to this problem, which is of interest to the Chinese government, is to develop tiger farms. The plan might involve transferring some of the revenue raised from the farms to help in the conservation of wild tigers.

Activity 3.8

What is your response to the idea of farming tigers? What do you think your response reveals about your own environmental values?

Comment

There are those who might find the idea of farming tigers an ingenious way to conserve the tiger population and those who might find the idea repugnant. The issue is complex, and you may well have changed your mind as you considered the various ways in which this development might be managed. But if you finally decided that tiger farming is acceptable, then you are suggesting that there is room for instrumental valuations and uses of tigers. At the same time you may have thought that this is a good way to protect wild tigers, thereby indicating your belief that the species' non-instrumental value is also important. You may, however, have decided that the tiger's non-instrumental value is paramount and that tiger farming is objectionable on the grounds that it degrades the animal. Both positions could be described as environmentalist, although the former could be described as close to what some call a 'shallow green' perspective, which is often focused upon using resources efficiently for long-term human use. The latter might be closer to a 'deeper green' perspective, which covers a range of attitudes and values but which tends to view environments and species as things that are not necessarily there for human purposes.

The tiger-farming question raises an issue that cannot be handled by the instrumental/non-instrumental distinction alone. We have been able to discuss how things are valued, but not how things perhaps should be valued, or indeed how we might reflect critically on our own values. To approach these issues, we need a second distinction that is frequently made in discussions of value: that of subjective versus objective.

Subjective and objective

You may prefer to eat jam rather than marmalade. However, you would not argue vehemently that your preferences were the right preferences. You may have a preference that tigers are farmed or not farmed, and here you might be more inclined to argue that your values are the right values and ones that should be adopted by others. The distinction here is partly between matters of preference and matters of principle. If someone accepted that a range of views on tiger farming was permissible then they would be suggesting that it was a matter of taste or preference. However, those who find tiger farming to be totally unacceptable, whatever the positive effects for wild tigers, would be arguing that farming tigers was wrong as a matter of principle and would be unlikely to agree to differ with others who held alternative views.

Another way of expressing the difference here is to use the term **subjective value** to apply to those cases where the values are considered to be a matter of taste or preference, and **objective value** to refer to those cases where there is a claim that something more than taste is at stake and there are wider principles involved. These labels also have another important aspect. If something is objectively valuable then its value is not thought to be solely dependent on the opinions, feelings or attitudes of sentient creatures (Rowlands, 2000). This is by no means an easy concept to grasp, nor is it one that is necessarily agreed upon (see Box 3.1), but by exploring this distinction you should develop your skills in analysing your own values and those of others. Activity 3.9 will help you to think about some of the issues at stake.

subjective value

objective value

Box 3.1   Differences between subjective, objective and intrinsic values

The idea that things can have value in and of themselves, independently of people or other creatures valuing them, has been attractive to some environmentalists. The term **intrinsic value** is often used to convey this sense that things have independent value. It is attractive because it sounds like a powerful statement to say that something is valuable whatever we may think about it. To some, this is more impressive than saying that it is valuable because a lot of people seem to like it. But the idea of totally independent value is highly contested and can seem a little unreasonable. How can value exist without there being someone or something to value it?

intrinsic value

In order to get around this problem we prefer the term objective value. Objective values refer to the values that people hold as principles and which they are prepared to argue with others about. In addition, it is sometimes useful to think of objective values as neither wholly dependent on, nor separate from, the people doing the valuing. The thing, or object, being valued has some say or part to play in its valuation even if, contrary to intrinsic value, it doesn't determine its own value. You should note that, like the instrumental/non-instrumental pairing, we are using this distinction to help you analyse how you and others value environments, and how we might consider valuing environments if we want to achieve ends like conservation of species or habitats. The distinctions are not set in stone and there will be grey areas. As Figure 3.12 indicates, they are as useful as you make them.

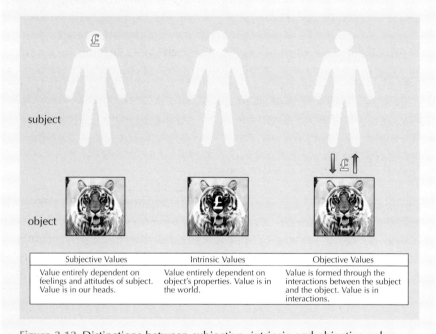

Subjective Values	Intrinsic Values	Objective Values
Value entirely dependent on feelings and attitudes of subject. Value is in our heads.	Value entirely dependent on object's properties. Value is in the world.	Value is formed through the interactions between the subject and the object. Value is in interactions.

Figure 3.12 Distinctions between subjective, intrinsic and objective value

Activity 3.9

Once your have decided if the following items are instrumentally or non-instrumentally valuable, try to decide if they are subjectively or objectively valuable. It may help to consider first whether it is worth arguing against someone who says that it is not valuable (if you are prepared to argue then it is more likely to be objectively valuable, or a matter of principle). It might also be useful to think about whether or not the object's value is solely dependent upon people's preferences. As we have said, this is more controversial, but you should consider the possibility that some of these things are thought to have value that is at least in part unrelated to people's tastes, feelings or attitudes.

1   a human heart

2   jewellery

3   Lake District National Park

Comment

A human heart is instrumentally valuable. It pumps blood around the body and so helps to keep tissues alive. It passes the substitute test as it could be replaced with a pig's heart or an artificial heart – although even here the case is a difficult one as some object to xenotransplantation. It is also objectively valuable. If someone told you that hearts didn't matter then you would argue that they were wrong, especially if they were actually going to act on their beliefs. In addition, it could be said that the heart is objectively valuable in that to make such a claim does not depend on our feelings or attitudes, or even whether or not we know it exists.

The second case, jewellery, tends to be thought of in different ways from a heart. While some may see jewellery as a means to an end, it is most often thought of as non-instrumentally valuable. However, just because it is often thought of as an end in itself does not mean that it is objectively valuable. Indeed, few would argue about jewellery as a matter of principle.

The Lake District and other National Parks in Britain are valued instrumentally, as spaces for recreation, but also non-instrumentally, for their beauty (Figure 3.13). Many people support organizations that work for their protection and resist attempts to build or quarry in such places. The same people urge others and especially children to visit and appreciate places like these. We could say therefore that these people value national parks objectively in that they are willing to argue for their existence both for their own use and for the use of others (and for future generations). It is possibly more debatable whether or not the parks have objective value in the stronger sense of not being entirely dependent for their value on people. This is something for you to think about.

You should now have a reasonable sense of two major distinctions that can be made with regard to values. Table 3.3 is a summary of these distinctions.

Table 3.3   Ways of distinguishing among values

Instrumental/non-instrumental	is the matter being valued as a means to something else or as an end in itself?
Subjective/objective	is the value a preference or are there principles at stake? Is the value solely dependent upon people's feelings and attitudes?

Figure 3.13 The Lake District National Park. Ought we to care about places like this?

The examples given in the above activity should make it clear that these distinctions are not necessarily related or overlapping. It is important to interrogate both distinctions when discussing value. We now turn to see how these can be applied to discussions of the value of those species that are threatened with extinction.

3.2 Why values matter: the example of extinction

In the previous chapter you were presented with evidence that species extinctions are currently occurring at a rate that is very possibly unprecedented. Our question here is what kinds of value, if any, are being lost when a species goes extinct? And, do we need to value non-human species in particular ways in order to help stem the tide of extinctions?

Extinction represents, first of all, the loss of things that have instrumental value. For example, there's a danger that a combination of overfishing and changes in sea temperatures will lead to the extinction of cod (*Gadus* spp.). If this happens an important source of food for humans will be lost. In addition, species like this either are, or could be, an important source of medicines, knowledge, materials and so on. Valuing something instrumentally may therefore give us *some* basis for arguing that a species should be saved. The same principle is often used for habitats and ecosystems. Tropical rainforests are often defended by virtue of the existing and potential uses to which the myriad of species contained in them can be put. We could also extend this to argue that recognizing the instrumental value of a species for other, non-human, species, is also important: we eat cod, but so do seals. As the previous chapter made clear, when any species is lost, members of some other species will possibly suffer as a result. If we care about their survival or well-being then we ought to care about the species on which they depend.

So, is instrumental value sufficient to ensure species survival? Unfortunately, while it may help to save some species, it is unlikely to help others. Valuing a species instrumentally, as food for humans, forms the basis of, at best, a weak case for the species' survival. The reason is simple: humans can provide their protein intake from many sources (and seals may well be able to survive on other fish). Even if we think there is little substitute for fresh cod, its existence in the wild may be given little or no value. Cod farms may mean that we have no qualms about the extinction of this species in the Atlantic. Like the question of tiger farms that you thought about earlier, while the species itself is saved, it is, you might agree, a fairly limited form of survival.

Instrumental values are also inadequate where few obvious interdependencies exist between species. While Atlantic cod might be missed by predators in the ocean, few if any species would miss the tiger. Tigers are predators that exist at the top of the food chain, meaning that very few, if any, other species are dependent upon them. If tigers are only valued instrumentally as part of TCM then there is very little hope for wild tigers. Finally, many more species that are not valued for something, or do not have an instrumental value, will be difficult to save if this is the only form of value that is considered legitimate.

Therefore, we may need to value some threatened species non-instrumentally if they are to survive (see Figure 3.14). They are not, in other words, valued solely as a resource or means to some other good, but are seen as an end in themselves. Tigers are one example of a species that is valued for its appearance. They, like some other species, are also valued for their rarity. Other species may be valued for their age or the complexity of their life-cycle and so on. However, relatively few species are valued in this way. Indeed, when compared to the vast number of species introduced in the previous chapter, their number is minute.

○   List some of the species that are commonly thought of as non-instrumentally valuable. What do they have in common?

●   Your list might include several large mammals, including whales, tigers, pandas, gorillas, some birds such as song thrushes and colourful insects such as the monarch butterfly. They all tend to be charismatic creatures.

It is relatively easy to view the list in the above question as subjective. The value of the species seems to depend almost solely on the feelings, attitudes and approval of human beings (that is, the values are a matter of preference rather than principle). While it would be wrong to belittle such feelings and attitudes, it is also reasonable to wonder how arbitrary this list can seem.

Figure 3.14 The Humpback whale (*Megaptoa noveangliae*). Should these animals be conserved because they provide goods that we can use, or should we think that they are worthy of preservation for their own sake?

The subjective nature of our values can become a liability when it comes to taking action to protect a species that is non-instrumentally valued. If we can only argue that it is our preference that a species is valued in this way then it can become extremely difficult to insist that resources are directed to the species' survival. Or we must rely on undemocratic and authoritarian means to further our cause. A preferable position would be to argue that there is just cause for valuing this species non-instrumentally and that our arguments have legitimacy beyond our own feelings and attitudes. In more simple terms, we would need to consider not only how a species is being valued but also the extent to which such a valuation is justified.

So, while it may be necessary, at least in part, to value species non-instrumentally, as ends, it is not necessarily sufficient to do so. We need to go further and ensure that the value of a species is not simply subjective. If we are to make strong cases for the survival of a wide range of non-human species that are not useful to us or to other creatures that we care about, or that are not simply charismatic, then we may need to call upon objective values.

We have used species extinction to discuss how value impinges on the survival, or otherwise, of certain species. The importance of values, however, is not limited to species. The question of what kinds of value are being used and what their effects are can be extended to any number of environmental questions, from the expansion of urban areas to atmospheric pollution. You should get into the habit of asking what kinds of value are in play, and what kind should be in play, when you consider the range of environmental issues in this series and elsewhere.

Summary

This section has discussed how we can distinguish among values and how different types of value affect environments (the second question posed in the Introduction). The main points of this section are:

- Species and environments can be valued either instrumentally and/or non-instrumentally.
- Species and environments can also be valued subjectively or objectively.
- Objective values are based on matters of principle. Subjective values are based on matters of preference.
- How something is valued can determine whether it is likely to be maintained or used to extinction.
- The example of species valuation suggests that a range of values are important in considering whether or not a species can be saved.

4   Who cares wins? The importance of inequalities and action

By now you should have a feeling for the importance of values in understanding environmental issues. We need to understand how environments are valued in order to understand the expectations and motivations that people bring to them. Having said this, it may not always be possible to convert values into action. For example, there may well be conflicts of interest that render some cherished principles redundant. Even where people stick to those principles, there is always the possibility that something or someone else, with a different set of concerns and interests, will be able to muster more power to successfully pursue their aims. This section focuses on these conflicts of interest. The main concern is to demonstrate how uneven distributions of power explain why some interests and values gain prominence and why some actions are more likely to succeed than others. Our case study is the Indian tiger (*Panthera tigris tigris*). Before we look at some of the power issues, there is a need to review the current plight of tigers, to understand the extent of the threats that they face.

4.1 Do Indian tigers need saving?

There are at least two elements to this question. First of all, we are asking whether or not tigers can survive without human intervention to protect them. In order to answer this question you will need to learn something about the current tiger population, its rates of decline, its spatial characteristics and the nature of the threats to tigers and tiger habitats. Some of the approaches that you learned about in the previous chapter will be useful in this respect. Second, the question can be

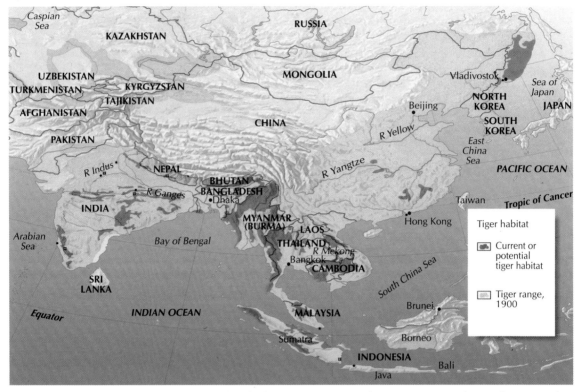

Figure 3.15 The range of tiger species one hundred years ago was much more extensive than is the case today. *Source*: Thapar, 1999.

read as asking whether or not wild tigers should be saved? Are they valuable and, if so, how are they valuable? This section is mostly concerned with the practical question of whether or not tigers can survive, but you should also reflect upon why some might think that tigers *should* survive as you study this material.

It is estimated that there were roughly 80,000 tigers at the start of the twentieth century, ranging over an area stretching from Vladivostock to the Black Sea (Seidensticker et al., 1999). This range is now reduced to a few scattered pockets of potential tiger-sustaining habitats (see Figure 3.15). The last century saw the extinction of two subspecies (the Balinese, *P. tigris balica*, and Caspian tiger, *P. tigris virgata*), and estimates now indicate that between 5,000 and 7,000 tigers exist in the wild (that's the total for all five existing subspecies: the Siberian, *P. tigris altaica*; the Indian, *P. tigris tigris*; the Corbetts, *P. tigris corbetti*; the Sumatran, *P. tigris sumatrae*; and the South Chinese, *P. tigris amoyensis*). Numbers continue to fall, possibly by several hundred each year (Thapar, 1999).

Roughly half to two-thirds of the total population of wild tigers live in India. The threat to this population stems from a number of actions. The hunting or poaching of tigers is one reason for the dramatic reduction in the total number of tigers in the wild (and indeed is a threat even to tigers held in zoos). Some estimate that one tiger is poached every 18 to 24 hours in India (although this figure is a sensitive one and is challenged by authorities: see Thapar, 1999). Wild

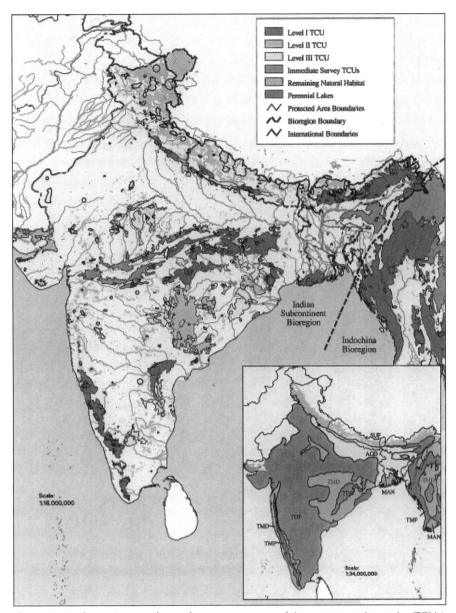

Figure 3.16 The main map shows the current extent of tiger conservation units (TCUs). A TCU is a block or cluster of blocks of habitat that contains, or has the potential to contain, interacting populations of tigers. The levels of TCUs (Level 1 being the highest) have been estimated following a survey of a number of factors including habitat integrity, poaching pressure and tiger abundance. Areas where survey data was not available, but where there is potential for tiger conservation, are classified as 'immediate survey TCUs'. The inset map shows the original or maximum extent of tiger habitat types. The tiger habitat types shown are: TMD – Tropical moist deciduous forests; TMF – Tropical moist evergreen forests; TDF – Tropical dry forests; AGD – Alluvial grasslands and subtropical moist deciduous forests; SUF – Subtropical and temperate upland forests; MAN – Mangroves.
Source: Wikramanayake et al., 1998.

tigers are often killed because they pose a threat to farmers and their livestock; however, more often they are killed for the commercial value of the tiger corpse. A tiger can fetch US$10,000 on the black market – this is equivalent to a lifetime's wage in some parts of India. Tiger pelts fetch large sums, but it is the use of tiger parts in TCM that accounts for most of a dead tiger's value.

While the hunting and poaching of tigers is a significant problem (and, to many, is the most distasteful threat to tigers), of greater overall significance is the depletion of tiger habitat. Figure 3.16 shows how tiger habitat has collapsed over the course of the last 100 years. The inset map depicts the maximum extent and range of habitats suitable for tigers. It is likely that the extent of these habitats has changed over geological time, but even with this long-term change in mind, comparison with the situation today suggests that something dramatic has happened in recent years.

The original range suggests that tigers are remarkably flexible in terms of the kinds of habitat in which they can thrive (from Mangrove swamps, to tropical moist evergreen forests, from tropical dry forests to grasslands). However, research on tiger populations suggests that they seem to prefer areas where grassland and forest form a landscape mosaic, and vegetation types vary over relatively small areas. These landscapes are favoured by the tiger's prey, which is entirely made up of large ungulate or 'hoofed' species: the sambar (*Cervus unicolor*), a type of Asiatic deer, and nilgai (*Boselaphus tragocamelus*, see Figure 3.17), or great Indian antelope, are favourite sources of food). Indeed, where this type of prey is abundant, tigers tend to prosper, and did so for a long period of human habitation of Asia, when small-scale, and often temporary, forest clearance produced the kind of landscape mosaic favoured by tigers and their prey.

Figure 3.17 Any human·activity that reduces the grazing opportunities for the nilgai (*Boselaphus tragocamelus*) will reduce the prospects for tigers.

However, over the course of the twentieth century, human hunting of ungulates and the human collection of forest products that would otherwise provide food for tiger prey, along with pressures on land for farming and other uses, have resulted in pressure on tiger populations.

Project Tiger

In the 1970s action was taken by the Indian Government, in association with international conservation organizations, to stem this decline of tigers and tiger habitats. The action was called Project Tiger and was part of a national parliamentary act, the Wildlife Protection Act, 1972. The Act as a whole was a legal framework designed to 'prevent the flagrant abuse of India's wildlife' (Thapar, 1997a, p.12). Project Tiger involved the designation of 23 tiger reserves with a combined area of 33,000 km^2 (see Figure 3.18 for locations). In the early years some success was recorded as tiger populations stabilized and even increased in number. However, the threats to tigers have not necessarily abated and in recent years matters may have become worse than ever. Human hunting of tigers and their prey remains a problem and, even in those areas designated as tiger reserves, there are numerous pressures to develop land for human purposes. Figure 3.18 includes a list of some of the land use changes that are threatening tiger reserves.

Activity 3.10

In what ways is land valuable, instrumentally or non-instrumentally, for the following?

(a) The developers of the projects listed in Figure 3.18

(b) The tigers

(c) Those people who want to save tigers by maintaining tiger reserves as areas that should be protected from the effects of large-scale developments?

Comment

Developers value land instrumentally, as a means for generating revenue from crops, energy projects or transportation links. The land is also valuable to tigers instrumentally as a productive food store. People who value wild tigers value the land that tigers require instrumentally, as a means to wild tiger conservation. Saving wild tigers requires maintaining the habitats and ecological relationships upon which tigers depend. It is also possible to discern a slightly more complex reasoning behind how group (c) might value the land. Conservationists can also value the land and the habitats supported by the land non-instrumentally – as things worthy of consideration on their own terms. In doing this they may value the tigers instrumentally as means to raise awareness, support, legal protection, funding or other resources that are vital in the battle to stave off what they regard as damaging development. Tigers can, then, be thought of as an umbrella species, under which other species and habitats can, in conservation terms, shelter.

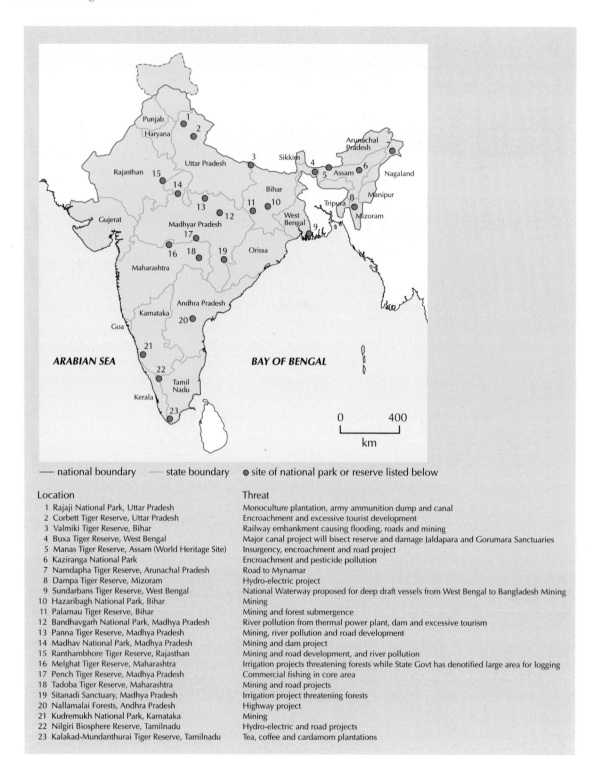

— national boundary   — state boundary   ● site of national park or reserve listed below

Location

1 Rajaji National Park, Uttar Pradesh
2 Corbett Tiger Reserve, Uttar Pradesh
3 Valmiki Tiger Reserve, Bihar
4 Buxa Tiger Reserve, West Bengal
5 Manas Tiger Reserve, Assam (World Heritage Site)
6 Kaziranga National Park
7 Namdapha Tiger Reserve, Arunachal Pradesh
8 Dampa Tiger Reserve, Mizoram
9 Sundarbans Tiger Reserve, West Bengal
10 Hazaribagh National Park, Bihar
11 Palamau Tiger Reserve, Bihar
12 Bandhavgarh National Park, Madhya Pradesh
13 Panna Tiger Reserve, Madhya Pradesh
14 Madhav National Park, Madhya Pradesh
15 Ranthambhore Tiger Reserve, Rajasthan
16 Melghat Tiger Reserve, Maharashtra
17 Pench Tiger Reserve, Madhya Pradesh
18 Tadoba Tiger Reserve, Maharashtra
19 Sitanadi Sanctuary, Madhya Pradesh
20 Nallamalai Forests, Andhra Pradesh
21 Kudremukh National Park, Karnataka
22 Nilgiri Biosphere Reserve, Tamilnadu
23 Kalakad-Mundanthurai Tiger Reserve, Tamilnadu

Threat

Monoculture plantation, army ammunition dump and canal
Encroachment and excessive tourist development
Railway embankment causing flooding, roads and mining
Major canal project will bisect reserve and damage Jaldapara and Gorumara Sanctuaries
Insurgency, encroachment and road project
Encroachment and pesticide pollution
Road to Mynamar
Hydro-electric project
National Waterway proposed for deep draft vessels from West Bengal to Bangladesh Mining
Mining
Mining and forest submergence
River pollution from thermal power plant, dam and excessive tourism
Mining, river pollution and road development
Mining and dam project
Mining and road development, and river pollution
Irrigation projects threatening forests while State Govt has denotified large area for logging
Commercial fishing in core area
Mining and road projects
Irrigation project threatening forests
Highway project
Mining
Hydro-electric and road projects
Tea, coffee and cardamom plantations

Figure 3.18 Land use conflicts in tiger habitats in India.
Source: Thapar, 1999

We have established that tigers are threatened both through hunting and through habitat depletion. The threat may now be even more serious, given the spatial distribution of the tiger populations. As we know from Chapter 2, small, isolated populations are especially vulnerable.

The consequences of such small and spatially isolated tiger populations are profound. The tiger population numbers only a few thousand, and the fragmentation of tiger conservation units together with large expanses of land that separate those units have resulted in the isolation of tiger populations. There are at least two important consequences. First, the lack of connectivity between tiger populations may result in inbreeding, genetic deterioration and therefore poor adaptability of the species. Second, the isolated habitat patches are often insufficient to sustain tiger populations for long periods. Habitats are dynamic and unstable and therefore prone to change. Any reduction in environmental quality (as viewed from a tiger's perspective), through human action, disease, floods or fires, can lead to local extinctions. These areas will not be re-colonized because all of the routes, or corridors, between tiger habitats have been lost. There are currently about 160 areas of tiger habitation world-wide, between which there is no dispersal or connectivity (Seidensticker et al., 1999, p.3). Even within these areas populations are fragmented. A prediction widely made is that extinction in more than half of these areas is almost inevitable.

So tiger populations *are* in danger of extinction. Whether or not they are valued sufficiently to warrant the resources needed to save them is a matter for you to consider, having read this and previous sections. But even if enough people agree that, in principle, tigers are worthy of being saved, there remains the question of what kinds of actions are needed and whether or not there is the power to take those actions.

4.2 Can tigers be saved? People, tigers and poverty

There are powerful forces that threaten tigers and their habitats. The list of threats includes poaching, hunting of tiger prey, reductions in habitat quality through collection of forest products, land use changes through large-scale development projects (roads, dams, mines, etc.), climate change, population fragmentation and reduction in genetic diversity of wild tigers. In order to consider whether or not tigers can be saved we need to appreciate how power works to either perpetuate or restrict these threats. A full analysis would need to be truly interdisciplinary, looking at all the threats and their interrelations. You already have some information on the importance of population size and fragmentation from Chapter Two. You can read more about the effect of land-use changes and climate changes on biodiversity in **Morris and Turner (2003a) and Morris (2003b)**; and **Blackmore and Barratt (2003)**. We can only be indicative here. The focus is on human land-use. Our aim is to highlight the importance of uneven social power and social inequality to environmental issues.

This will help us to add further detail to our understanding of the ways in which power enables and constrains action.

Most of the tiger reserves created under Project Tiger include or are surrounded by villages with significant human populations (Mackinnon et al., 1997). All of them are in rural India where people mostly make their living from farming or by directly collecting resources from the land (including gathering wild fruits and nuts, and hunting). In effect this means that people and tigers are often reliant on the same ecological relationships or on parts of the same food webs (see Chapter Two for a definition of food webs). There is, therefore, a potential conflict between people and tigers.

○   Note the phrase 'potential conflict'. How might some of these activities benefit tigers?

●   It was noted earlier that some human-induced changes to habitats can produce a landscape mosaic which provides more opportunities for tiger prey to flourish and so indirectly helps tigers to survive.

How, then, has a situation that was once beneficial to both tigers and people become so conflictual? One answer is that India's human population has risen so rapidly that, at over 1,000 million people, it is now extremely difficult to provide enough land and resources for both tigers and people. However, this is not the whole story and may not be the critical issue. Indeed, when considering the effects of human populations on other species and more generally on environments it is necessary to consider not only the number of humans but also the kinds of activities that people engage in, and the conditions within which they find themselves. There are two major questions to ask of a situation in which environments are threatened. Have people's activities changed? Have the conditions in which people live altered?

Before we continue with tigers we shall try to apply these to the passenger pigeon case. In brief, we would suggest that in terms of the first question it was the rise of a passenger pigeon industry and, for the second question, the growth in urban consumption, that best explain the bird's sudden decline.

In rural India the scene may be somewhat more complex. In terms of the first question it is true to say that there are many new activities that are having effects on tigers. Dam building and other large-scale projects are testament to the growing power of some sectors of Indian society that can mobilize domestic and foreign capital to change landscapes and livelihoods in rural areas. This is a powerful movement in India, one that has grown since the mid-1980s and is centred on a new economic policy. In general terms this policy involves reducing barriers to trade and to international business and the privatization of previously state-held assets. There are some powerful interests at work. International business interests are being effectively melded to government interests. While the former can find new profits in new markets, the latter can increase revenue from exports, can secure loans from the International Monetary Fund (IMF) and the World Bank, and can use this influx of finance to reduce deficits. These interests can mobilize vast resources in terms of money, expertise and technology.

There are also discourses at work. Once such discourse is **globalization**.   globalization
Globalization is a term used to convey the sense that people and events are increasingly connected through activities like trade, media and environmental processes. The term is explained in greater detail in **Castree (2003)**. Globalisation, along with modernization and development of rural India, are described as inevitable by government and business. India's urban elite dismisses any resistance to this kind of development as idealistic or unrealistic. Instrumental values win out over any lingering non-instrumental valuation of landscapes or tigers. This results in significant changes in rural activities, many of which could be detrimental to the tiger's survival.

This is not the whole story because tigers are also under threat in those places where large projects are absent. There are many people in rural India who are not benefiting from these projects. For these people, rural life is still concerned with cultivating the land, grazing livestock and so on. As we have seen, in the past these actions may not have caused conflict between people and tigers and may even have been of benefit to tigers. Yet tigers and tiger habitats are under threat even where there has been no obvious change in human activities. To understand why, we need to turn to the second question: have the conditions in which people live altered?

The general indicators of Indian living conditions suggest an increased standard of living over the last few decades. For example, the Indian economy grew at an average rate of 5.7 per cent per year in the 1990s. A growing economy is usually a good indication of increasing incomes and a healthy economic scene in terms of investment and development. The average rate of growth during the 1970s was 3.5 per cent, so the new figure is seen as evidence of an improvement and a justification for the new economic policy that was mentioned above. However, this does not tell the whole story. First of all, the growth has not been even. The agricultural sector, which employs the majority of people in India, has grown much more slowly (2.7 per cent). Meanwhile, the percentage of people who are described as below the poverty line has remained static in the 1990s. As the human population has risen over this period this can only mean that in absolute terms (i.e. in terms of actual numbers) the number of people enduring conditions of poverty has increased. In general terms, the new economic policy 'has favoured urban over rural populations, richer over poorer states and professionals over wage earners and the poor' (Vanaik, 2001, p.51).

Conditions may have worsened for the rural poor in India. It should be added that around tiger reserves this has sometimes been made even more problematic. Tiger reserves were designed to exclude people. For farmers this meant having to find new land to cultivate or graze. People were also cut off from collecting forest products. In many cases this led to intense environmental pressure being placed on land surrounding tiger reserves, resulting in poorer yields and eventually contributing to increased poverty. We now need to establish what this increase in production of poverty has to do with environmental issues in general and tiger survival in particular.

The passenger pigeon example suggested that growing wealth can increase people's ability to alter environments. As people assemble goods and resources so their capacity to act over ever-larger swathes of time and space grows. So why is poverty an environmental issue? In the case of the tiger, one stark answer to this question is that poverty makes the $10,000 available for a tiger carcass more than just a passing temptation. In other words, poverty and the power relations that produce it can have effects on values. It is idealistic (and self-righteous) to expect that people living in conditions of poverty will all value environments or species non-instrumentally.

Poverty also indirectly affects tigers. It is more likely that people will use tiger lands and forests for the illegal collection of food and for grazing animals if they are living in conditions of hunger (Figure 3.19). This depletes the food stock for tiger prey. Poverty may also mean that the infrequent loss of cattle to a roaming tiger no longer becomes tolerable, so herdspeople hunt down the offending individual, and possibly kill indiscriminately. Other practical considerations relate to available funds. In rural areas where finances are scarce, there is little money to police reserves or protect tigers from poachers. Likewise, within a population

Figure 3.19 Bihar, in rural India. Many millions of people live on the margins of survival. Poverty can be produced by environmental problems and can also provoke human action to further degrade the environment. To address environmental problems, it is often necessary to address matters of social inequality.

there are sometimes problem tigers that need careful and expensive management in order to prevent attacks on people and livestock.

Returning to the question of whether tigers can be saved, you should now have a working knowledge of some of the issues at stake. The question is closely tied to power. We have argued that powerful groups are building up resources in India and internationally in order to develop land in particular ways: ways that are in their interests but may be detrimental to India's environments generally and tigers in particular. Doubtless, those powerful groups will be keen to suggest that economic growth helps to build a better environment for India's people and wildlife. While there might be some truth in this, we should remember that this build-up of power and economic wealth is not necessarily shared equally. This uneven distribution of power is an important way of understanding the context within which powerful groups operate. It is far easier to assemble resources and build up power when those who might oppose you are living under duress. So not only is it necessary to think about the resources of power, and the discourses that help to make that power legitimate or seem inevitable, it is also vital to look at the power relationships between groups. If some are becoming more powerful at the expense of others, or at the same time as others are becoming impoverished or are failing to make themselves heard, then the effects of power can be all the more evident. These effects start to have environmental implications if the powerful fail to consider environmental matters as important, and/or if those who are increasingly powerless start to become impoverished and lack the resources to live sustainably in their environments. We bring together some of these issues in Activity 3.11.

Summary

Section 4 has concentrated upon the interaction between power and successful action with regard to the Indian tiger population. From your reading of this section you should have a better idea of how power works to enable some and constrain others in their response to environments; our third question posed in the Introduction. The following activity should help you work through some ideas about power and action

Activity 3.11

Outline what changes, if any, you would make to the Project Tiger system of tiger reserves in order to promote the effective protection of tigers. It may be useful to consider the spatial characteristics of the reserves and the uneven distribution of power and wealth that contributes to their success or otherwise.

Comment

This is a live issue with many possible answers that you may want to discuss with others. When we thought about it we noted that spatially the reserves have been isolated, internally fragmented and have supposedly had tight boundaries limiting people's access. It may make sense in the future to think about ways of linking tiger habitats together by tiger-friendly land-use corridors in order to boost the chances of small populations surviving. Meanwhile, Project Tiger reserves may have done little to reduce conflicts between people and tigers. Indeed, if they have contributed to rural poverty then it may be necessary to think of reserves as more permeable, allowing people access to some, though not all, tiger lands in order to farm. This policy would need to be promoted in conjunction with attempts to reduce rural poverty and explore ways of people and tigers cohabiting more successfully. Reducing poverty requires us to look carefully at the current distribution of power in India, and also between India and the rest of the world. If elites continue to favour large international projects and if little is done to improve living conditions in rural areas then the future of the rural poor and tigers may be bleak.

These final points underline the difficulties in taking concerted action regarding the conservation of tigers. Taking such action requires pressure being placed on international businesses and on governments to consider issues of poverty and environmental value as matters of vital importance. If living, wild tigers are not seen as valuable there will be little action to save them. Meanwhile, if relieving poverty is not seen as an important aspect of economic development then the environmental effects could be disastrous, and the results are unlikely to be sustainable.

5   Conclusion

The primary aim of this chapter has been to demonstrate the relevance of values, power and action to environmental issues. The chapter has demonstrated that human actions have become increasingly responsible for environmental changes and that to understand those actions, and possibly change them, we need to understand people's motivation and their capabilities for action. In discussing motivations a number of useful distinctions between environmental values have been made. You should now be able to recognize these distinctions in the case studies we have looked at and in other areas of environmental concern. It should also be possible for you to reflect on your own values, and consider the extent to which you value aspects of environments.

You should now be able to describe how some people and groups enjoy more power to change their environments than was perhaps the case in the past, and to understand how this power is built up by such groups in a variety of situations. You should also be able to discuss the importance of discourse in enabling some forms of environmental action and inhibiting others. We have also introduced the importance of inequality as a driver of environmental change. You should be able to understand why environmental projects depend on improving the conditions for human as well as non-human inhabitants.

Finally, you should be able to describe and explain how values, power and action are related to one another. Broadly, what people do, or do not do, relates to what they want to see happen as well as to how they understand their ability to make things happen. The importance of this to understanding everyday actions, as well as those actions that are more consciously directed to bringing about change, should be something that you can apply to other people's situations as well as your own.

References

Blackmore, R. and Barratt, R.S. (2003) 'Dynamic atmosphere: changing climate and air quality' in Morris, et al. (eds).

Bingham, N., Blowers, A.T. and Belshaw, C.D. (eds) (2003) *Contested Environments*, Chichester, John Wiley & Sons/The Open University (Book 3 in this series).

Castree, N.C. (2003) 'Uneven development, globalization and environmental change' in Morris, R.M. et al. (eds).

Cronon, W. (1991) *Nature's Metropolis: Chicago and the Great West*, New York, Norton.

Eales, S. (1991) *Earthtoons: The First Book of Eco-humour*, New York, Warner Books Inc.

Hemley, G. and Mills, J.A. (1999) 'The beginning and the end of tigers in trade?', in Seidensticker, J., et al. (eds).

Mackinnon, K., Mishra, H. and Mott, J. (1997) 'Reconciling the need of conservation and local communities: global environment facility support for tiger conservation in India' in Seidensticker, J. et al. (eds).

Maples, W.E. (2003) 'Environmental justice and the environmental justice movement' in Bingham, N. et al. (eds).

Morris, R.M., Freeland, J.R., Hinchliffe, S.J. and Smith, S.G. (eds) (2003) *Changing Environments*, Chichester, John Wiley & sons/The Open University (Book 2 in this series).

Morris, R.M. and Turner, C. (2003a) 'Dynamic Earth: processes of change' in Morris, R.M. et al. (eds).

Morris, R.M. (2003b) 'Changing land' in Morris, R.M. et al. (eds).

Ponting, C. (1991) *A Green History of the World*, London, Sinclair Stevenson.

Price, J. (1999) *Flight Maps: Adventures with Nature in Modern America*, New York, Basic Books.

Reddish, A. (2003) 'Dynamic Earth: human impacts' in Morris, R.M. et al. (eds).

Rowlands, M. (2000) *The Environmental Crisis: Understanding the Value of Nature*, London, Macmillan.

Seidensticker, J., Christie, S. and Jackson, P. (eds) (1999) *Riding the Tiger*, Cambridge, Cambridge University Press.

Thapar, V. (1999) 'The tragedy of the Indian Tiger: starting from scratch' in Seidensticker, J. et al. (eds).

Vanaik, A. (2001) 'The New Indian Right', *New Left Review*, vol.9, pp.43–67.

Wikramanayake. E.D., Dinerstein, E., Robinson, J.G., Karanth, U., Rabinowitz, A., Olson, D., Mathew, T., Hedao, P., Conner, M., Hemley, G., and Bolze. D. (1998) 'An ecology based method for defining priorities for large mammal conservation: the tiger as a case study', *Conservation Biology*, vol.12, no.4, pp.865–78.

Chapter Four

What to do? How risk and uncertainty affect environmental responses

Nick Bingham and Roger Blackmore

Contents

1   Introduction: what to do?

Early in the previous chapter you encountered an image of a vast flock of passenger pigeons blacking out the sky over nineteenth-century America. From the fate of the birds – driven to extinction as a species in the face of massive commercial hunting – you would have appreciated why such historical drawings are the only way that we can, today, get a visual sense of this at once terrifying and exhilarating sight. Thinking through this situation using the concepts of values, power and action you were able to develop an analysis of why the extinction happened, and how to transfer the lessons of this example to other environmental situations.

In order to outline the objectives of this chapter we begin with a comparable image, this time comprised of millions of individuals of a species of butterfly commonly known as the monarch (*Danaus plexippus*). Like the sight of the passenger pigeon flock, a spectacle like this one provoked feelings of both fear and wonder when experienced at the close of the nineteenth century. For example, in 1892 the appearance of such a mass of monarchs over the US city of Cleveland, Ohio, alarmed many of the people who saw it, with a local newspaper describing it as a 'living, breathing, palpitating picture' (Grace, 1997, p.39). Unlike the sight of the passenger pigeon flock, however, we still have the opportunity today of witnessing a swarm of migrating monarch butterflies (of which Figure 4.1 is a recent photograph). Nevertheless, the future of this migration – which has been called 'the most striking and spectacular butterfly phenomenon on earth' (Wells et al., 1993, p.463) – is by no means certain. As we shall see in Section 2, the monarch faces threats at critical points along its journey. In this context, the most urgent issue is not how this situation has arisen (although that is certainly important in terms of lessons for the future), but what

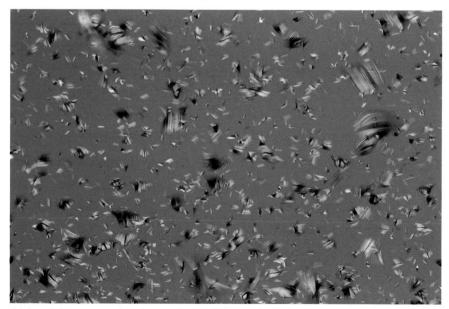

Figure 4.1 A swarm of migrating monarch butterflies.

sorts of responses we should make to it. As we shall see, both in terms of the monarch and of environmental issues more generally, the question of what to do is far from easy to decide. What this chapter will show is how the analytical concepts of risk and uncertainty can help guide us through such complexity.

The main aim of this chapter, then, is to provide a final response to one of this book's key questions: how can we make sense of environmental issues? In Chapter Two, we saw that one response was to look at physical and biological changes and to suggest why, where and at what rates these changes are occurring. This approach was summarized in the key concepts of time and space. A second response, discussed in Chapter Three, was to investigate why certain environmental objects and issues become important to certain individuals and groups, and how the different resources available to those individuals and groups can affect the outcomes of contests over those objects and issues. This approach was summarized in the key concepts of values, power and action. In this chapter, we will focus on a third way of getting to grips with environments, through the key concepts of risk and uncertainty.

In order to do this, we will be taking a slightly different tack from the one adopted in the previous two chapters; in those chapters key concepts were defined and developed within the text, and you were shown how applying them to environmental issues of various kinds could illuminate those issues for you. Here, we are faced with a situation in which risk and uncertainty are not just organizing principles that are useful for studying environmental issues, but have come to play a central role in influencing how environments are treated much more generally.

Since the late 1960s there has been a marked shift in how the world's industrial nations have approached issues of protecting environments: increasingly, there has been a shared recognition that if environmental policy is to be effective it has to concern itself far more with preventing future harm and avoiding potential problems, rather than merely dealing with harm after it has already occurred. The results of this shift in environmental priorities can be seen in many arenas; from policy documents and party manifestos, through regulations and laws, to government action plans and international treaties (as well as in the practices of both private and public sector organizations), environmental thinking has become much more future oriented (for more on this shift, see Jasanoff, 1999).

This is where the concepts of risk and uncertainty come in. In environmental matters at a policy level, just as in our everyday lives, actions are almost always taken in the light of imperfect, incomplete or unclear information. Short of the sun rising tomorrow and taxes being levied, there is not much we can be truly confident about. In order to help establish what we can be more – or less – confident about concerning environmental issues, approaches have been developed that use ideas about risk and uncertainty to deal with what we do not yet know. For example (and as we shall see in Section 4), regulatory bodies that are responsible for implementing environmental laws are increasingly required to justify their decisions on the basis of formalized risk assessments. In turn, those responsible for creating a new product or process are increasingly obliged to satisfy the regulatory bodies that they have taken appropriate steps to measure and minimize any risks either to human health or environments that may be associated with it.

| hazards |
| probabilities |
| unknowns |
| attitudes |
| risk assessments |
| indeterminacy |
| trust |

Although there is some sort of consensus today that risk and uncertainty are central to environmental responses, there is much less agreement on precisely what those concepts mean and how they should be used. As we shall see in this chapter, when people talk about risk, uncertainty and environments they can be talking about some related but very different things, including the seven we will explore: hazards, probability, unknowns, attitudes to risk, risk assessment, indeterminacy and trust. Our task therefore is not so much to provide definitions of our own, but to give you a sense of this range of different usages. By the end of the chapter you should feel confident that you can both distinguish between them and recognize why there might be differences of opinion about which particular version of risk and uncertainty is the most appropriate in a given situation.

2   Dangerous environments?

The *first* thing that people talk about with respect to environmental risks and uncertainties is danger of some sort. This has not always been the case. Although at the time of its emergence in Europe during the Middle Ages the concept of risk was associated with uncontrollable events such as storms, floods or epidemics (often seen as 'acts of God') and the disruption to human life associated with these, by the nineteenth century it had developed more neutral connotations. By then, risk and uncertainty could refer to either 'good' or 'bad' outcomes, gains or losses. At the start of the twenty-first century, this distinction has been lost to some extent (although it is still present in business where risk takers are often positively valued) and risk and uncertainty are once more primarily related to negative or undesirable outcomes, not positive ones (for more on the history of risk, see Lupton, 1999).

2.1 Hazards

hazard

| hazards |
| probabilities |
| unknowns |
| attitudes |
| risk assessments |
| indeterminacy |
| trust |

As you saw in Chapter One the danger to be found in relation to environments is usually treated as being embodied in hazards of some sort. When we are considering environmental issues, a **hazard** may be broadly defined as any event, process or product that poses a threat to humans, non-humans or their habitats. Sometimes a distinction is made between 'natural' hazards, such as floods, and 'human-caused' hazards, such as pesticide overuse, depending on what or who is being blamed. However, in practice it is very unusual to encounter a hazard that is only natural or only social in character (Hinchliffe, 2000). In order to get a sense of what specifically can count as a hazard in environmental situations, it will be helpful now to revisit the theme of extinction and endangerment that we have been developing over the last two chapters. In particular, we can return to the case of the monarch butterfly.

2.2 A threatened phenomenon

Although the monarch butterfly is globally widespread, the North American population of the monarch is unique in being predictably migratory over very large distances. This migration has its origins in the distant past but for our purposes what you need to know is that there are currently two migratory populations of the monarch in North America (see Figure 4.2).

One population breeds west of the Rocky Mountains, and in the summer the butterflies migrate south-westwards to stay over winter at low altitude in the forested sites along the Pacific Coast of California. A much larger population breeds east of the Rocky Mountains and migrates in the autumn up to 3,600 km during about 75 days (about 50 km per day) to over-winter in high-altitude forests in Central Mexico. Year after year, in both populations the autumn migrants use the same highly localized areas to stay in over the winter months, despite the fact

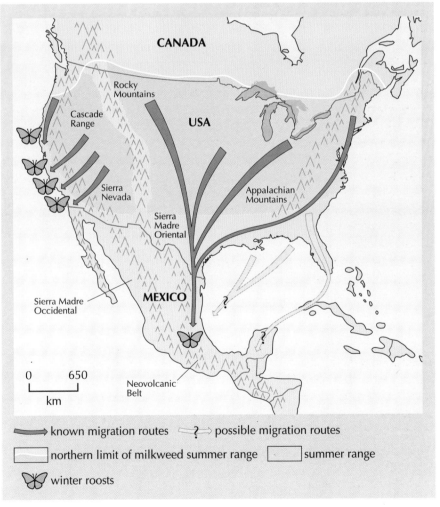

Figure 4.2 Many and far: the monarch's North American migration.
Source: adapted from Grace, 1997.

that the individuals that move south are between three and four generations away from their ancestors that occupied the sites in the previous winter.

The precise mechanism that determines how the monarchs are able to make this remarkable journey with such unerring accuracy remains a mystery, but we do now have a much better sense of why they take it in the first place. Following the discovery in 1975 of the exact location of the over-wintering grounds in Michoacan state, Mexico, ecologists and biologists have been able to ascertain more precisely why such a limited area was used over so many generations of butterflies. (It is worth noting that the notion of 'discovery' involved here really means revelation to the outside world, since local peoples had known for many centuries that the monarchs arrived around what they called the time of 'the day of the dead', bringing back the souls of departed children.)

It had long been assumed that the migration was driven by the need to find warmer winter conditions than were available in most of the USA and Canada. However, a straightforward need for warmth turned out to be a far from adequate explanation of the monarchs' choice of winter habitat. Certainly, if it is too cold they freeze in numbers far greater than the population can withstand if it is to survive. On the other hand, if it is too warm they tend to fly more, thereby burning up the stored fat that they need to fuel themselves through the winter and on the return flight in the spring. In order to survive the winter, then, the wanderers require an extremely narrow and highly specific set of circumstances.

The colonies, such as those shown in Figure 4.3, form at altitudes of about 3,000 metres above sea level, usually in stands of oyamel. Oyamel is a native fir that the butterflies seem to prefer as a roost, as they can cling to the fine, needle-covered branches in great numbers. The sites tend to be on southern or eastern slopes, exposed to the warmth of the midday or afternoon sun. They also have low undergrowth that allows any butterflies falling from the trees at temperatures too cold to allow flight to crawl up and away from the ground, where temperatures drop to near freezing at night and in the early morning. Studies have shown that this delicate balance may be upset by any factor that opens up the protective forest canopy to the harsher influences of wind, rain, snow and cold by night, and to intense sunshine, desiccation and overheating by day.

Figure 4.3 Tucked up tight: monarch butterflies roosting over-winter in a fir tree, Mexico.

The migration of the North American monarch population, however, is being threatened at various sites along its route. According to leading authorities the combined effects of these threats may lead to the loss of the migration (and thus of the population itself)

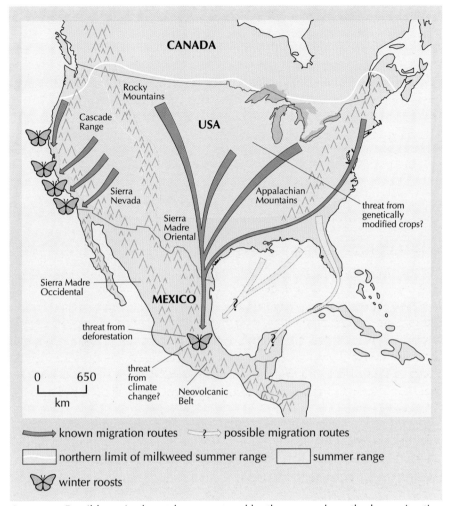

Figure 4.4 Possible major hazards encountered by the monarch on the long migration routes.
Source: adapted from Grace, 1997.

in just the next few decades. As you can see from Figure 4.4, the North American monarch faces three potential hazards in particular: deforestation, genetically modified crops and climate change.

Deforestation

The most important hazard is that posed by deforestation. Almost as soon as the Mexican over-wintering roosts were discovered in the mid 1970s, there were reports that a combination of logging and land clearance was having a devastating effect on these areas and their future ability to support monarch populations. These actions were in turn caused by a complex set of social, political and economic factors, including poverty, changes in land ownership, clashes between the interests of indigenous peoples and peasant farmers, and increased demand from the USA for cheap timber of the sort found in these areas. Although at the time these reports were essentially observational and anecdotal, they were taken

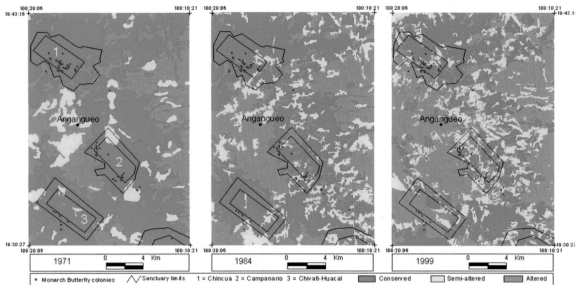

Figure 4.5 Satellite images of Michoacan forests at time intervals over thirty years. The images show changes in forest cover from 1971 to 1984 to 1999 in Central Mexico in an area that includes three of the twelve known massifs that serve as over-wintering areas for the monarch butterfly. Three categories of forest cover are shown: conserved forest cover, semi-altered forest cover and altered forest cover. The concentric black areas delineated are the core (internal) and buffer (external) zones. Red dots indicate all recorded locations of over-wintering colonies from January 1977 through March 1998; they do not represent colonies that form each year. *Source*: Brower et al., 2002.

increasingly seriously as evidence of forest loss began to accumulate. Today they have been fully substantiated by satellite images that show just how great the forest loss has been (see Figure 4.5).

Genetically modified crops

genetically modified (GM) crop

The second, and most controversial of the three, is the hazard that may be posed to monarchs by **genetically modified (GM) crops**. GM crops are crops that have been engineered to contain a gene or genes not normally found in that species of plant. As we shall see in more detail in Section 4, it has been suggested that certain types of corn that have been engineered to kill a particular pest are also toxic to monarchs in their caterpillar stage.

Climate change

The third is the hazard that may soon be posed to monarchs by climate change. You will probably be aware of the widespread concern that human activity has been accelerating cyclical patterns in global temperatures, particularly over the last couple of centuries or so (see **Blackmore and Barratt, 2003**). At present the major concern is that a marked warming is taking place. This increase in overall temperatures is likely to have many different consequences over many different timescales for humans, non-humans and their habitats. According to a report published in 2000 by the World Wide Fund for Nature (WWF, 2000), among the first to feel the effects will be species like the monarch that rely on isolated mountainous ecosystems for survival.

Activity 4.1

You now have a good idea of what constitutes a hazard for the monarch butterfly. To make sure that you can extend this recognition to other environmental situations, go back through the previous three chapters of this book and identify an event, process or product that represents a potential or proven hazard for (a) a human population, (b) a non-human species and (c) a particular habitat.

Comment

There are obviously many possible answers to this activity. We identified: emission of radiation from Bradwell nuclear power station or the proposed waste disposal site as a potential hazard to the nearby human population (Chapter One); hunting as a proven hazard to the North American passenger pigeon population (Chapter Three); and the sea level rise associated with climate change as a potential hazard to the marshland habitats around the Blackwater estuary (Chapter One).

Whatever your answer you will have had to be precise about a particular event, process or product being a hazard for a particular human population, non-human species or habitat. This is important because what may act as a hazard in one case may not in another. In other words there are specific features of specific human populations, non-human species and habitats that increase their potential to experience certain things as hazards. We can refer to these features as vulnerabilities.

Let's return now to the monarch.

○   From what you have read so far in this chapter, do you think that the monarch's situation in Mexico could be described as vulnerable?

●   Given what you already know about how entangled the fates of species and their habitats are, and what you have learned here about the specialized set of conditions monarchs require to survive over-wintering in enough numbers to ensure the next stage of the migration is viable, we would expect you to answer yes to this question. You would be right to assume that in the absence of similar habitats in the region, the dependence of the migrating monarchs on such a unique combination of circumstances does in theory expose them to considerable threats if something were to happen to the oyamel forests.

We can draw two other general conclusions about environmental hazards from the brief survey above. The first is that hazards are not always easy to pin down. We have mentioned deforestation and climate change as potential hazards so far in this chapter. While both are obviously major processes, they have only recently been identified as posing a threat to the monarch. Even now we have only an incomplete sense of what exactly causes these hazards in the first place. The other general conclusion is that hazards may be of different sorts. The two processes, deforestation and climate change, are the outcomes of many small natural and social actions. The possible hazard posed by GM crops to the monarch is related to the introduction of a particular innovation.

Summary

- Hazards are one of the things that people are often referring to when they are talking about environmental risks and uncertainties.
- Hazards can be defined as events, processes or products that can pose a threat to humans, non-humans or habitats.
- Hazards are related to particular vulnerabilities and may be of different kinds and are not always easy to identify precisely.

3   Manageable environments?

A key feature of the shift towards prediction in matters of environmental policy that we identified in the introduction has been the increasing quantification of environmental issues. Put simply, what this means is that decision makers have tried to find ways of measuring in numbers as many aspects of those issues as possible, from the populations of particular species to the volume and activity rates of releases of nuclear radiation. What is appealing about this quantification of environments is that they can be clearly presented, assessed and – very importantly – compared. For those attempting to manage environments in various ways, the ability to assess and weigh up different risks and uncertainties on a common (mathematical) scale is obviously potentially a very effective one, especially if probability can be assigned to outcomes. However, environments are complicated and we are often faced with unknowns rather than certainties or probabilities. This, in turn, makes the attitudes of those involved in decision-making processes central to what courses of action are followed. In this section we examine how these three things – probabilities, unknowns and attitudes – have become central to the future of our own environments and those of others.

3.1 Probabilities

| hazards |
| probabilities |
| unknowns |
| attitudes |
| risk assessments |
| indeterminacy |
| trust |

After hazards, then, the *second* thing to consider with respect to environmental risks and uncertainties is probability. To explore why in more detail, let's look at an example of the move to place numbers on the environmental processes that we have just described. Once again our choice will be guided by the ongoing theme of extinction and endangerment. As you should recall from the previous chapter, there may be many reasons – and many different kinds of reasons – why individuals and groups might find it important to protect particular species from extinction or endangerment. When considering butterflies, for example, we might encounter arguments based on both instrumental values (that highlight perhaps the work that butterflies do in fertilizing many cultivated plants) and on non-instrumental values (more concerned probably with butterflies' beauty or even their possible intrinsic right to exist). Butterflies, however, are only one class of what you now know are the huge range of species that may be counted as endangered in one form or another (and the monarch only one species within that class).

Given that the resources available for conservation work are limited, there is obviously a need to rank species according to the threats they face. Now, there is never any guarantee that the species likely to go extinct first will be prioritized as the most important species to save. However, it is true to say that such ranks and the categories on which they are based can play a crucial role in raising general awareness of problems and in setting priorities and targets for management action at the local, national and international scales.

One classification in particular, which – despite having no legal status – has become globally accepted and thus one of the most important tools in conservation circles, is that produced by the IUCN (the World Conservation Union or International Union for the Conservation of Nature and Natural Resources as it was previously known). The set of criteria adopted by that organization in the early 1970s, summarized in Table 4.1, became the norm for over twenty years when assessing and assigning a category of threat to most species. In 1994 when it produced a new classification that placed a much greater emphasis on a numerical approach, summarized in Table 4.2, it was a major event in the field.

Table 4.1   Summary of pre-1994 IUCN categories

Old IUCN categories	
EX	Extinct in your country or region. The species has not been located in the wild during the past 50 years.
E	Endangered. Species in danger of extinction and whose survival is unlikely if the causal factors continue operating. Include butterflies whose numbers have been reduced to a critical level or whose habitats have been so drastically reduced that they are deemed to be in immediate danger of extinction.
V	Vulnerable. Species believed likely to move into the 'Endangered' category in the near future if the causal factors continue operating. Include butterflies of which most or all the populations are decreasing because of over-exploitation, extensive destruction of habitat or other environmental disturbance.
R	Rare. Species with small populations that are not at present 'Endangered' or 'Vulnerable', but at risk. Mostly these species are localised within restricted geographical areas or habitats or are thinly scattered over a more extensive range.
I	Intermediate. Species known to be 'Endangered', Vulnerable' or 'Rare' but where there is not enough information to say which of the three categories is appropriate.
K	Insufficiently known. Species that are suspected but not definitely known to belong to any of the above categories, because of lack of information.

Source: www.IUCN.org

Table 4.2   Summary of post-1994 IUCN categories

IUCN categories

CRITICALLY ENDANGERED

A   Population reduction of at least 80% over the last 10 years.

B   Extent of occurrence less than 100 km^2 and two of the following:

　　1   severely fragmented or known to exist at only a single location;

　　2   continuing decline;

　　3   extreme fluctuations.

C   Population estimates less than 250 mature individuals and a strong decrease.

D   Population estimate less than 50 individuals.

E   Probability of extinction at least 50% within 10 years.

ENDANGERED

A   Population reduction of at least 50% over the last 10 years.

B   Extent of occurrence less than 5000 km^2 and two of the following:

　　1   severely fragmented or known to exist at no more than five locations;

　　2   continuing decline;

　　3   extreme fluctuations.

C   Population estimates less than 2500 mature individuals and a decrease.

D   Population estimate less than 250 individuals.

E   Probability of extinction at least 20% within 20 years.

VULNERABLE

A   Population reduction of at least 20% over the last 10 years.

B   Extent of occurrence less than 20000 km^2 and two of the following:

　　1   severely fragmented or known to exist at no more than ten locations;

　　2   continuing decline;

　　3   extreme fluctuations.

C   Population estimates less than 10000 mature individuals and a decrease.

D   Population estimate less than 1000 individuals.

E   Probability of extinction at least 10% within 100 years.

LOWER RISK

Three subcategories:

1   Conservation dependent (cd). Taxa on the focus of conservation programmes, the cessation of which would result in qualification for one of the threatened categories above a period of five years.

2   Near threatened (nt). Taxa not qualifying for Conservation dependent but close to qualifying for vulnerable.

3   Least concern. Taxa not qualifying for Conservation dependent or Near Threatened.

DATA DEFICIENT

There is inadequate information to make an assessment of extinction risk based on distribution or population status.

Source: www.IUCN.org

○   Comparing the two sets of categories and bearing in mind the comments that began this section, put yourself in the position of someone working in conservation and ask yourself the following questions:

(a)   What do you think were the strong points of the pre-1994 classification that helped it survive for so long?

(b)   Why do you think the shift to the new classification might have been made?

●   As we saw it, for a practitioner the simplicity of the earlier classification, the modest amount of data one would have needed to make a designation and the wide acceptance of the categories would certainly have been attractive features. That said, we also felt that perhaps a weakness might have been that there were no explicit guidelines for assigning species to categories of risk. Certainly one of the reasons the IUCN overhauled the existing system was because the previous classification relied so much on expert judgement it was difficult to achieve consistency among workers using it. The result was that when conflicting opinions did arise there was no systematic means of resolving the problem. We also reflected that in terms of getting results like convincing a governmental or non-governmental organization of the need to implement legal protection of a particular species, being able to back up a categorization with data of a numerical nature might be a powerful tool.

One of the innovations of the new classification system produced by the IUCN was that it became possible for the first time to represent the threat to a given species in terms of a calculation of risk.

As we noted at the outset of this chapter, risk and uncertainty can have many different meanings when used in relation to environments. When we say here that risk calculations were possible within the new IUCN criteria, we are indicating that risk is being used in its technical sense. You might recall that you were given a definition of risk in its technical sense in Chapter One of this book. There, it was emphasized that two things need to be known before making a calculation of risk. The first is the possible outcomes of a particular event. The second is the likelihoods of these outcomes occurring. Thus – to use an example from that chapter – if we wanted to know the risks (in a technical sense) associated with building a sea wall, we would require information about what might happen, for example saltmarsh loss, and how likely it is that saltmarsh loss will occur.

We have already spent some time discussing outcomes in the form of hazards. Before we move on it is also important that you have a basic sense of how likelihoods are worked out and for this you need to know about probability. Like risk, although we tend to use it relatively vaguely in everyday speech, it is important to recognize that probability has a specific technical meaning as well. To clarify exactly what that is, you should work through Box 4.1, which explores the term in some detail.

Box 4.1 Probability

Probability (p) can be defined numerically as the chance of a particular event or outcome occurring and can take any value on a scale from zero to one. If an event can never happen or has never been known to happen, the probability is zero, or $p = 0$; if the event is certain and always happens, then the probability is one, or $p = 1$. Probabilities are normally given in the form of a decimal, 0.5 for example, but they can equally well be expressed as a percentage (50 per cent) or fraction (½). Probabilities can be determined theoretically or by conducting a large number of trials.

If all the possible outcomes of a situation and the probabilities of each are known then we can begin to make predictions about what will actually happen. In the familiar example of tossing a coin the outcomes are either a head (H) or a tail (T). Assuming the coin is fair there is an equal chance of throwing either, and each has a probability of 0.5. What does this tell us? It cannot predict whether you will throw a head or a tail next time, but it does tell you that if you make a very large number of throws you can expect to get approximately equal numbers of heads and tails.

Two other features of this example should be noted. First, with a single coin you can either throw a head or a tail, not both; the two outcomes are *mutually exclusive*. Second, and less obviously, each throw of the coin is *independent*; that is, the result of the previous throw has no influence on the outcome of the present throw. If you have just thrown three heads in a row, it doesn't mean that a tail is now more likely to come up. As long as the coin has not been tampered with, the chance of either event occurring remains ½.

Consider now what can happen if we toss two coins together. We can draw up a table to show there are four possible outcomes. These outcomes are equally likely (you could check this for yourself with a trial of, say, 100 throws).

		Coin 2	
		H	**T**
Coin 1	**H**	H H	H T
	T	T H	T T

We can use these results to find the probability of getting 0, 1 and 2 heads respectively.

- No heads (i.e. two tails), TT, occurs only once out of four possible outcomes.

- One head (and one tail) HT, TH, occurs twice, or two outcomes out of four. (We are not concerned here with the order in which a head occurs.)
- Two heads, HH, also occurs once.

So, the probabilities of getting 0, 1 and 2 heads, based on the table, are 1/4, 2/4 (or 1/2) and 1/4 respectively, or more usually 0.25, 0.5 and 0.25.

In this example we have made use of two general rules of probability, which are summarized below.

1   Given two mutually exclusive events, with a probability of occurrence p_1 and p_2, the probability of one *or* other event occurring is $p_1 + p_2$.

   For example, the

   probability of occurrence of two heads = p_1 = 0.25

   probability of occurrence of one head = p_2 = 0.5

   So the probabaility of *either* two heads (p_1) *or* one head (p_2) = 0.25 + 0.5 = 0.75

2   The probability of two independent events occurring is $p_1 \times p_2$.

   For example, the

   probability of throwing a head in the first trial = p_1 = 0.5

   probability of throwing a head in the second trial = p_2 = 0.5

   So the probability of throwing a head in the first trial (p_1) and in the second trial (p_2) = 0.5 $\times$ 0.5 = 0.25

A useful rule of thumb is that if, for the two events, *either* one *or* the other is defined to be a successful outcome, then rule 1 applies; but if the first *and* the second throw must occur for the experiment to be a success, then rule 2 applies.

A third rule is also useful:

3   The expected number of occurrences of any event is $p \times N$, where p is the probability and N is the total number of occasions on which it could possibly occur (e.g. number of throws).

Activity 4.2

The following questions will allow you to test your understanding of the basic rules in Box 4.1.

(Hint: for each question decide which of the three rules applies.)

(a) How many times would you expect to get two heads from a trial of 50 throws of 2 coins?

(b) Suppose that on average, it rains once a week, so the probability that it will rain tomorrow is 1/7. What is the probability of it raining on either Saturday or Sunday (but not both)?

(c) What is the probability of getting two fours on throwing two dice?

Comment

(a) (Rule 3) The expected value is 0.25×50 or 12.5. This is the theoretical value. Clearly you can't get half a result, but in practice if you carried out many trials of 50 throws you would find 12 and 13 pairs of heads occurring most often.

(b) (Rule 1) The two events are mutually exclusive: either it rains on Saturday or on Sunday, so it is the sum of their probabilities: $1/7 + 1/7 = 2/7$ or 0.29.

(c) (Rule 2) The probability of getting a four with any (fair) dice is 1/6 (the die has six faces), the outcome of the toss of the second die is an independent event, so rule 2 applies and the probability of both dice showing fours is $1/6 \times 1/6$, that is, 1/36 or approximately 0.028.

Returning to the IUCN case that we have been following here, the outcome we are interested in is clearly the extinction of the species within a particular period of time. The likelihood is the probability of that extinction actually happening within that timeframe. In other contexts we could be talking about the chance of any of the other environmental hazards that we reviewed in Section 2 occurring. Whatever the particular situation, in terms of the moves to quantify environmental threats outlined above, being able to use risk in its technical sense is very attractive. Although knowing the probability of an outcome is not the same as predicting it, having a solid sense of its likelihood potentially makes a hazard much more manageable. Depending on the magnitude of the threat posed and the chance that it will take place over a given period, a hazard for which the risk is known can be planned for or ignored as is deemed appropriate.

In order to make sure that you have understood the basic idea of risk in its technical sense, try Activity 4.3.

Activity 4.3

Refer back to Table 4.2 and see if you can identify what risk a species would have to be facing if it was to be considered 'critically endangered'.

Comment

If you went to the right place, which was section E under 'Critically Endangered', you would hopefully have worked out that the criteria of risk would be a probability of 'at least 50 per cent within 10 years' of an outcome of 'extinction'.

This discussion has shown that if we know the probabilities of the possible outcomes of a particular event we can make a technical calculation of risk with respect to that event. Let's see what happens if we don't have so much information.

3.2 Unknowns

Being able to make a calculation of risk in a given situation is a very powerful tool, and one that environmental decision makers value highly. However, it is rare that enough robust information about both outcomes and probabilities exists for risk in its technical sense to be applicable to many environmental situations. There may be several reasons for this, including: a simple lack of available data; problems in measuring a given variable; the fact that no precedent exists for certain new hazards; or the complicating background of normal variability. An example should make the sorts of difficulties involved clearer.

We used the monarch butterfly earlier in this chapter to illustrate what sort of things could count as hazards in environmental situations and noted that the deforestation of its over-wintering roosts was the threat of most concern. In 1983 this concern resulted in the monarch migration being designated a 'threatened phenomenon' by the IUCN and the number one priority in world butterfly conservation. You may have noticed that this designation does not correspond directly with any of the pre-1994 categories of threat listed in Table 4.1. This is because at that time it was only entire species that could be categorized and – for all its problems in the North American context – the monarch is far from endangered as a species worldwide. In order to draw attention to the plight of the monarch's migration, then, the IUCN had to create a special label appropriate to the case.

Under the new classification, particular geographically limited populations as well as entire species can be assessed and so, in theory, the monarch's status could be reviewed. Although it has not been reviewed at the time of writing this chapter, from what you have learned so far in this section you should be able to recognize that perhaps the most crucial information one would need would be accurate information about past and present population numbers. Traditionally, because of the huge numbers and dispersion of the North American monarch population, obtaining such information has been carried out when as many as possible are all in the same place at the same time (i.e. when they are located in their Mexican roosts). By measuring the amount of treetop space taken up by the over-wintering monarchs in their five main sites and then reckoning that there are about 13 million monarchs per hectare in the roosts, some totals that are comparable from year to year have been generated (see Table 4.3 and Figure 4.6).

Table 4.3   Estimates of the size of over-wintering monarch population in Mexico 1994–2001

Year	Estimated size
1994/1995	101,153,000
1995/1996	162,630,000
1996/1997	232,310,000
1997/1998	41,340,000
1998,1999	67,730,000
1999/2000	77,610,000
2000/2001	29,770,000

Note: Figure 4.6 presents the data in this table graphically.
Source: Journey North, www.learner.org/jnorth

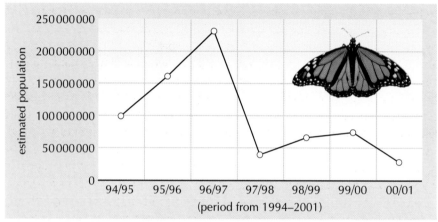

Figure 4.6 Estimated size of over-wintering monarch population in Mexico 1994–2001 (graphical representation of the data in Table 4.3).

Activity 4.4

Based on the numerical and graphical information on over-wintering monarchs in Table 4.3 and Figure 4.6, see if you can identify any broad trends in the population during that period. Do you think that we can make any judgements about threats to the monarch from this data?

The response to this acitvity is discussed in the text that follows.

From the data in the table and figure, we can see that in terms of trends, although the population at the end of the series (2000/2001) is less than half of that at the beginning (1994/1995), it clearly cannot be said that there has been a steady decrease in numbers over that period. On the contrary, between the first and third winters there was a significant increase up to a peak, followed by a sharp decline to a level that remained relatively stable over the final four winters. In this case, then, the available information does not indicate any obvious patterns. However, even if it did, we would have to be very careful about making any judgements about a connection with the hazards posed to the monarch by deforestation. On the one hand the data on which the table and figure are based are of precisely the sort that would be needed for any full IUCN-style analysis of threat. On the other hand, in order to ascertain how much of any trend is indicated, we would need to separate this data from the degree of numerical fluctuations or habitat alteration that could be expected under normal (i.e. no deforestation) circumstances. In other words, it is only after establishing the background 'noise' associated with an environmental situation that the impact of more drastic effects often associated with human actions can be truly calculated.

In the case of the monarch, this background noise can be quite variable. All butterfly populations vary widely in abundance from year to year due to differences in birth rate and death rate influenced by changing environmental factors (see **Drake and Freeland, 2003,** for more on population dynamics). A particular species may have a characteristic long-term abundance such that we might term it either 'common' or 'rare', but experience considerable variation about this average level from generation to generation. For example, if food supplies are good (with abundant flowers producing nutritious nectar for butterflies) then their reproductive output (the number of fertile eggs laid and offspring produced) may be increased. If foods for the larvae are then plentiful and predators and parasites scarce, subsequent caterpillar survival may be high (warm temperatures can further hasten development of all these stages). The opposite can also apply of course and the general lesson is that the supply of resources and habitat conditions and the behaviour of other organisms present in the ecosystem inevitably produce variations in the total numbers of any butterfly species and in the amount of change (up or down) between any two succeeding generations. As is often the case with migratory species, the monarch displays a particularly extreme version of this variation. It has been estimated that total numbers can vary by a factor of 10 over time, and in the past both the eastern and western monarch populations have suffered losses approaching 90 per cent (particularly following severe winter storms) and still recovered when the surviving 10 per cent experienced optimal conditions in the breeding range.

In practical terms, what this means is that until a reliable method is found of distinguishing between the normal variability of monarch populations and any reduction attributable to deforestation, it will be impossible to calculate with any confidence the risk of the future extinction of that population. Although the outcome we are working with is clear, the probability remains unknown. Unknowns are the *third* in our list of things that people tend to talk about with respect to environmental risks and uncertainties. More precisely, we can talk in terms of uncertainties. Just as risk has both a general sense and a technical definition, so too does uncertainty. We have seen that risk in a technical sense can be defined as something for which we need to know both possible outcomes and the probabilities of those outcomes. **Uncertainty** in a technical sense can be defined as the conditions in which we operate when we know the possible outcomes but cannot assign a probability to those values.

hazards
probabilities
unknowns
attitudes
risk assessments
indeterminacy
trust

uncertainty

We noted above that there are not that many environmental situations in which we can make a meaningful risk calculation in a technical sense. Much more common, however, are circumstances of uncertainty. Our monarch conservation case is one relatively small-scale example. Deciding how to respond to possible climate change represents the other extreme. If separating the normal from the anthropogenic (human-caused) changes poses such problems for one species, imagine how difficult it is to do for a phenomenon such as the climate that varies so profoundly over such long timescales (for more on how this is done, see **Blackmore and Barratt, 2003**). Part of the answer is to use mathematical ways of taking account of and representing uncertainties, and the IUCN recommends using two of these – best estimates and ranges of possible values – as ways of dealing with measurement error and normal variability. Beyond these, however, the way that conditions of uncertainty are handled in environmental issues is very much dependent on the attitude to risk of the relevant decision-making institution, which is the focus of our next section.

3.3 Attitudes

| hazards |
| probabilities |
| unknowns |
| attitudes |
| risk assessments |
| indeterminacy |
| trust |

Attitudes to risk are the *fourth* thing that people tend to talk about with respect to environmental risks and uncertainties. Helpfully for our analysis here, the IUCN has a very well defined attitude to risk, usually called precautionary. In terms of applying the criteria of classification, what this approach means is that the IUCN recommend that a species should (within sensible limits) be categorized as threatened unless there is clear evidence that it is not so. As you may recall from Chapter One, however, the precautionary attitude to risk has a much wider relevance within environmental matters. In fact it has become – particularly in the more familiar form of the 'precautionary principle' – one of the most important guides to policy making and action in situations of uncertainty.

Something like prevention by precaution has been operating for a long time in fields such as medicine and public health, where the old adage 'better to be safe than sorry' is often applied. However, the adoption of an explicit and coherent concept called the precautionary principle as relevant to environmental hazards and their uncertainties only began in earnest during the 1970s. At that time German scientists and policy makers were trying to formulate appropriate responses to the phenomenon of 'forest death' (*Waldsterben*) and its causes, notably air pollution. The main outcome of their deliberations was a rule of public policy action that could be applied generally in situations wherever there are potentially serious or irreversible threats to human health or environments. That rule can be summarized as the imperative to take action before there is absolute proof of harm (taking into account the likely costs and benefits of action and inaction). Since the 1970s, the precautionary principle has become an increasingly important part of the environmental agenda, becoming enshrined – albeit in different ways – in numerous international agreements (see Table 4.4). Note that although the words are similar, the precautionary principle is only one part of the broader preventative approach to environmental harm that we identified in the introduction to this chapter as characterizing recent policy making.

Table 4.4   The precautionary principle incorporated into various international treaties and agreements

The 'precautionary principle' in some international treaties and agreements

Montreal Protocol on Substances that Deplete the Ozone Layer, 1987

'Parties to this protocol ... determined to protect the ozone layer by taking precautionary measures to control equitably total global emissions of substances that deplete it.'

Third North Sea Conference, 1990

'The participants ... will continue to apply the precautionary principle, that is to take action to avoid potentially damaging impacts of substances that are persistent, toxic, and liable to bioaccumulate even where there is no scientific evidence to prove a causal link between emissions and effects.'

The Rio Declaration on Environment and Development, 1992

'In order to protect the environment the Precautionary Approach shall be widely applied by states according to their capabilities. Where there are threats of serious or irreversible damage, lack of full scientific certainty shall not be used as a reason for postponing cost-effective measures to prevent environmental degradation.'

Framework Convention on Climate Change, 1992

'The parties should take precautionary measures to anticipate, prevent or minimise the causes of climate change and mitigate its adverse effects. Where there are threats of serious or irreversible damage, lack of full scientific certainty should not be used as a reason for postponing such measures, taking into account that policies and measures to deal with climate change should be cost-effective so as to ensure global benefits at the lowest possible cost.'

Treaty on European Union (Maastricht Treaty), 1992

'Community policy on the environment ... shall be based on the precautionary principle and on the principles that preventive actions should be taken, that the environmental damage should as a priority be rectified at source and that the polluter should pay.'

Cartagena Protocol on Biosafety, 2000

'In accordance with the precautionary approach the objective of this Protocol is to contribute to ensuring an adequate level of protection in the field of the safe transfer, handling and use of living modified organisms resulting from modern biotechnology that may have adverse effects on the conservation and sustainable use of biological diversity, taking also into account risks to human health, and specifically focusing on transboundary movements.'

Stockholm Convention on Persistent Organic Pollutants (POPs), 2001

Precaution, including transparency and public participation, is operationalized throughout the treaty, with explicit references in the preamble, objective, provisions for adding POPs and determination of best available technologies. The objective states: 'Mindful of the Precautionary Approach as set forth in Principle 15 of the Rio Declaration on Environment and Development, the objective of this Convention is to protect human health and the environment from persistent organic pollutants.'

Source: EEA, 2001.

For the IUCN, the history of conservation has proved that acting in a precautionary manner is necessary in order to get results. For example, we have seen that the IUCN assigned to the monarch migration its 'threatened phenomenon' classification when there were many uncertainties about the situation. Without that early action, it is unlikely that the steps toward the protection of the over-wintering areas, which have been taken since that designation, would have been made. These began with a decree by the President of Mexico in 1986 creating a Monarch Butterfly Special Biosphere Reserve of 16,000 hectares designed to protect the most threatened of the roosts, and progressed since then towards a new, larger, and much better managed reserve of about 56,000 hectares designed to provide a sustainable solution to the problem.

Summary

In this section we have used conservation as an example of the broad shift towards prevention in environmental policy making identified in the introduction. We have seen how even in this limited context environmental risks and uncertainties can be and are considered in at least three ways:

- *Related to probabilities*
  If we know the probabilities of the outcomes of a particular event we can make a technical calculation of risk with respect to that event. Being able to calculate risk in this technical sense is very highly prized by environmental policy makers as it allows comparison between hazards.

- *Related to unknowns*
  Because of their social and natural complexity, most environmental situations do not lend themselves to being calculated in terms of risk in a technical sense. Much more often we are faced with unknowns of various sorts, and they characterize contexts when possible outcomes are known but probabilities cannot be assigned to those outcomes.

- *Related to attitudes*
  An example of an attitude to risk often applied to environmental issues is precaution, which – in the form of the precautionary principle – has become part of many major environmental agreements.

4   Unpredictable environments?

So far in this chapter we have reviewed four of the seven different ways that we identified in the introduction as ways in which people tend to talk about environmental risks and uncertainties. In order to introduce some basic concepts we have deliberately chosen cases studies that (i) involve hazards that are relatively well understood and (ii) are relatively uncontroversial in terms of responses. Although, as we saw, considerable uncertainties about the precise effects of deforestation on monarch populations still exist, the basic ecological mechanisms at work are now becoming increasingly known. At the same time (all other things being equal) we would expect that most people would support the work of an organization like the IUCN or want to protect at least some of the monarch's vital over-wintering grounds. In the final section of the chapter the hazards we look at are different on both counts.

First we will turn our attention to the hazards associated with the introduction of new technologies. These are often poorly understood and are increasingly a concern for environmental policy makers. Whether we are thinking in terms, for example, of wind-powered turbines (**Blowers and Elliott, 2003**) or biotechnologies, predicting the unintended environmental consequences of innovations has become one of the most important challenges facing environmental policy. Second, and as a result, we will be encountering situations where there are disagreements between different groups on what are the most appropriate ways of dealing with the risks and uncertainties. These disagreements can have significant effects on how policy responses are developed, as we discover next when we revisit the theme of extinction and endangerment and the case of our monarch butterfly.

4.1 The 'national insect' and GM

The monarch – certainly a charismatic species as far as insects are concerned – represents an excellent example of the process whereby for various reasons some species become more valued than others by certain individuals and groups. This is true to such an extent that the butterfly has been adopted by a number of US states (including Illinois, Vermont, Texas, Idaho and Alabama) since 1975 as their state butterfly or insect. Looking at Figure 4.7, it is not difficult to see why. The butterfly is big, beautiful and, as we have already seen, widespread. There has even been a campaign to have it officially recognized as the country's national insect. Although this has not yet been successful, the image and symbolism of the monarch have become almost ubiquitous in the USA as far as environmental and particularly conservation issues are concerned.

It was therefore something of a shock when in 1999 a short piece published in the scientific journal *Nature* raised the possibility that genetically modified crops could have adverse effects on the monarch. It was doubly shocking, since these same crops had been portrayed by the corporations who had developed and introduced them as both technologically advanced and environmentally friendly.

Figure 4.7 A single monarch: one of the most popular insects in the USA.

We can trace the background to this possibility of adverse effects to 1962 and the publication of *Silent Spring*, Rachel Carson's hugely influential work describing the environmental harm done by many chemical pesticides. A direct result was that some agricultural companies began to look for alternatives. One biologically based approach that was explored was to manipulate the soil organism *Bacillus thuringiensis* (Bt) which secretes a protein that, when ingested by a sensitive insect, causes the larval gut to break down and death to follow. Industrial and academic scientists have since selected a number of strains of Bt that are toxic to larva of different insects in order to produce 'natural' or bio-pesticides. With the advent of the new science of genetic engineering, which makes it possible to transfer genes between different species of animal and plant, a new method of utilizing these features of Bt became possible. When successful, the gene transfer enabled by these techniques can give the recipient species some of the traits found in the donor species. An obvious strategy, therefore, was to insert various Bt genes into crop plants in the hope that the genetically modified plant would display the same toxicity to pests as Bt did.

Agricultural companies introduced the Bt genes into several crops, including potatoes, soybeans, cotton and corn. One major target of this effort was the European corn borer moth (*Ostrinia nubilalis*) shown in Figure 4.8, a species that can damage these crops to an economically significant extent throughout the eastern USA and southern Canada. Before any of these genetically modified organisms could be grown commercially, the Environmental Protection Agency (EPA) of the USA required a number of tests to be done on the safety of these new plants. The toxins of various Bt strains showed no apparent adverse effects on honeybees, ladybirds and a few other invertebrates. These test results, along with the fact that the toxin is rendered harmless in the stomach of humans and other mammals, led the regulatory agencies to judge nearly all Bt corn strains safe for human consumption and for the environment generally.

See **Bingham (2003)** for more on GM as an environmental issue.

Then, in May 1999 an article was published by researchers at Cornell University who had been investigating whether pollen from Bt corn plants could accumulate on plants that grow extensively in and adjacent to cornfields and thus inadvertently kill (in the same way as conventional insecticides) native insects that are not pests. To test this hypothesis, they chose the monarch as their non-target species. Female monarchs lay eggs on wild milkweed plants, the only plants their caterpillars can eat (see Figure 4.9). In their experiment, conducted in a laboratory, the authors dusted the leaves of the common

Figure 4.8 The European corn borer moth (*Ostrinia nubilalis*), which lives up to its name causing extensive economic damage to many crops in North America.

milkweed with pollen gathered from one of the Bt corn strains. They found that caterpillars that fed on the dusted leaves ate less, grew more slowly and suffered higher mortality than caterpillars reared on milkweed leaves dusted with pollen from a non-BT corn strain. The scientists were guarded about their results and stated clearly that more research was needed to determine the impact of the toxic pollen on the monarchs in their natural

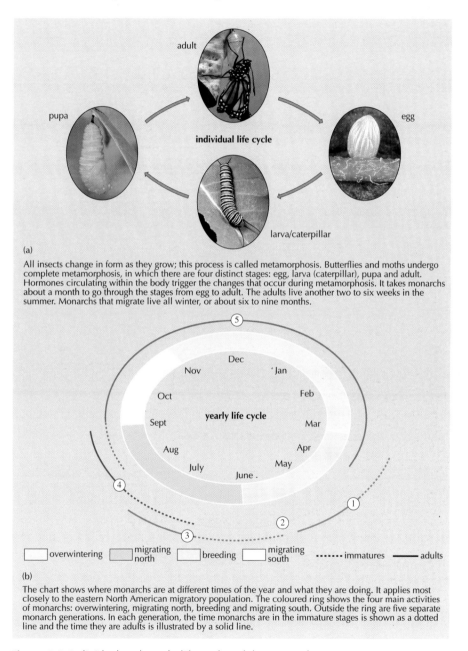

(a)

All insects change in form as they grow; this process is called metamorphosis. Butterflies and moths undergo complete metamorphosis, in which there are four distinct stages: egg, larva (caterpillar), pupa and adult. Hormones circulating within the body trigger the changes that occur during metamorphosis. It takes monarchs about a month to go through the stages from egg to adult. The adults live another two to six weeks in the summer. Monarchs that migrate live all winter, or about six to nine months.

(b)

The chart shows where monarchs are at different times of the year and what they are doing. It applies most closely to the eastern North American migratory population. The coloured ring shows the four main activities of monarchs: overwintering, migrating north, breeding and migrating south. Outside the ring are five separate monarch generations. In each generation, the time monarchs are in the immature stages is shown as a dotted line and the time they are adults is illustrated by a solid line.

Figure 4.9 Individual and yearly life cycles of the monarch.
Source: adapted from *Basic Biology: Yearly Life Cycle*, www.monarchlab.umn.edu/ BBio/ylc.htm

environment. Despite this call for caution, however, the response on publication was anything but, and a national outcry followed.

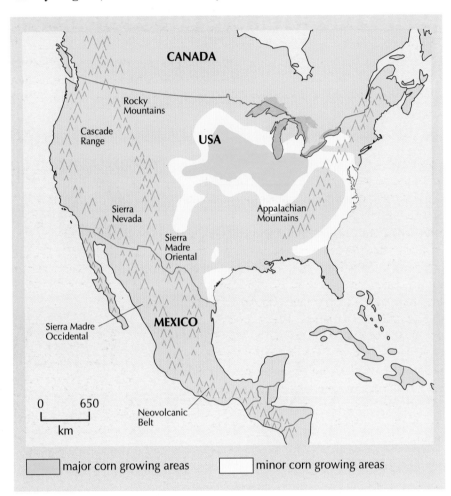

Figure 4.10 The main corn growing areas of the USA.

Activity 4.5

Looking at Figure 4.10 and reflecting on the previous sections, why do you think the reaction to this research might have been so marked?

Comment

One reason was clearly the existing perception of the monarch as a threatened species in the USA. Another was probably awareness that a significant proportion of the migrating monarch population (Figure 4.2) would in any given year pass through precisely the areas where corn (including GM Bt corn) is mainly farmed, as shown in Figure 4.10. Finally, although not nearly as intense as the controversy in Europe (see **Bingham, 2003**), support in the USA for GM foodstuffs was by no means total. Altogether, it is not difficult to see how even the suggestion of a connection between monarch deaths and GM crops acted to both crystallize existing concerns and create others. To underline just how strongly this connection did take hold of the public imagination, note the use of the image of the monarch in both Figures 4.11 and 4.12, which are publicity materials from two (very different) organizations – the Union of Concerned Scientists and Keep Nature Natural – both involved in the debate on GM food that was sparked by the *Nature* piece.

Faced with this situation the question was the one with which we started this chapter: what to do? Two very different responses to this question were made by two very different groups involved with environments. Both groups based these responses on the same two sources of available scientific information. The first source was comprised of existing literature on the medium- to long-term behaviour of monarchs in existing (non-GM) cornfield ecosystems. The second was made up of necessarily more short-term and small-scale laboratory or field studies of interactions between monarchs and Bt corn in particular. What divided the groups then was not data, but – as we will see next – an incompatibility with respect to the ways in which risk and uncertainty were handled.

Figure 4.11 Publicity material of the Union of Concerned Scientists with prominent monarch logo.

Figure 4.12 Publicity material of Keep Nature Natural with prominent monarch logo.

4.2 Risk assessments

The first group, whose response to the possible toxicity of Bt corn to monarch larvae we shall trace, is the EPA. As the body responsible for granting the biotechnology companies the licence to grow Bt corn commercially in the first place, the EPA came in for a great deal of criticism when the *Nature* story broke. In particular it was accused of relying too heavily on data provided by the biotechnology companies (in whose commercial interests it was to have the crops licensed) in order to make its earlier decision, and ignoring at that stage the potential impact on species of butterflies and moths that were not the target of the Bt trait but shared the same ecosystem in which the corn is grown. Stung by suggestions that it had allowed an unsafe product onto the market, the EPA spent the period from 1999 to 2001 undertaking a major information gathering exercise designed to both review and supersede its original decision to permit the release of Bt crops. In effect, the agency was carrying out a whole new risk assessment.

| hazards |
| probabilities |
| unknowns |
| attitudes |
| risk assessments |
| indeterminacy |
| trust |

risk assessment

Risk assessments are the *fifth* thing that people tend to talk about with respect to environmental risks and uncertainties. **Risk assessments** are an integral part of the preventative approach to environmental harm, which – as we saw in the introduction to the chapter – guides current environmental policy. Basically, their purpose is to find out what is known about a particular, potentially hazardous situation and the options for dealing with it. Although there are many different variations on the basic risk assessment procedure, these three main stages are common to most:

- *Risk identification* is the attempt to isolate and characterize all the possible outcomes from choosing a particular option.

- *Risk estimation* uses analytical methods to estimate the probability of each outcome and the magnitude of the adverse effects associated with that outcome (see Section 3.1 above). It is often considered to be the more objective or science-based stage but the assertion that this is also value free will usually be contested.

- *Risk evaluation* (which involves the decision maker) uses this technical information, together with any additional relevant information, to evaluate the alternative actions available. Risk evaluation is concerned with judging the significance and acceptability of risks, explicitly involves value judgements and should include consideration of risk perception and risk benefit studies.

The particular variant used by the US EPA is known as comparative risk assessment (CRA) which combines more technical and more judgement-based analyses. Once a problem area or issue of concern has been identified a set of risk categories is selected that typically includes ecological risk, human health risk and some measure of social risk. The initial ranking of risk within each set, referred to as 'risk assessment', is undertaken by groups of experts in individual areas; a second 'risk management' stage of attempting to reconcile the rankings related to risk types using a range of factors is the task for a group of stakeholders, often community-based groups. The overall aim is to group problem areas into priority categories.

In the Bt case the deadline for which the new risk assessment had to be carried out was the end of September 2001. This was what the EPA had set – previous to the monarch controversy – as the date by which it would make public a decision on whether or not to renew the licences to grow GM Bt crops in the light of their safety record since they were first approved. On 16 October 2001, after a final delay to allow the inclusion in the assessment of a crucial set of studies on the ecological implications of the crops on the monarch, the EPA released a press notice announcing their decision.

It stated that 'corn genetically modified with *Bacillus thuringiensis* (Bt) has been approved for an additional seven years' and that 'scientific studies and a history of successful use have demonstrated that Bt is not toxic to humans or other animals'. Also noted was that:

of particular concern during this [review] process were the potential risks to Monarch butterflies. In investigating these risks, the Agency requested extensive data from the scientific community in order to better evaluate the potential concern. The scientific evidence demonstrates that Bt corn does not impact Monarch butterfly populations. EPA has also determined that there will be no effects to endangered species from the use of the currently registered crops.

(EPA, 2001)

New requirements for farmers were introduced by the EPA designed to aid in 'strengthening insect resistance management, collecting research data on potential environmental effects, and improving grower education and steward-ship'. However, overall, the EPA message was made very clear. The benefits that Bt crops offer in terms of an 'effective, low-risk pest control alternative, which helps to protect the environment by reducing the amount of conventional pesticides used', far outweighed any worries about safety (EPA, 2001).

The conclusive nature of the language used by the agency and the scale of the review exercise it undertook seemed to indicate that as far as the monarch was concerned, the Bt crops issue was now closed. However, as we have already indicated, not everyone was happy with how risk and uncertainty had been dealt with by the EPA.

4.3 Indeterminacy

The second group whose response to the monarch and Bt controversy that we want to follow is actually a collection of leading environmental non-governmental organizations (**NGOs**), including the Science and Environmental Health Network (SEHN), Greenpeace, and the Union of Concerned Scientists (UCS). (An NGO is defined by Willets, 2002, as 'an independent voluntary association of people acting together on a continuous basis, for some common purpose, other than achieving government office, making money or illegal activities'.)

Before, during and following the re-registration of Bt crops in late 2001, they all expressed severe reservations about the way in which the EPA was treating the hazards potentially posed to the monarch by Bt corn. At its most basic, their criticism was that by using conventional risk assessment methods as described above, the EPA was displaying an attitude to risk and uncertainty that was inappropriate for dealing with new technological innovations such as GM crops. More than that, statements like 'the scientific evidence demon-strates that Bt corn does not impact Monarch butterfly populations' contained in the EPA press release quoted earlier were felt to be misleading. For the NGOs such claims imply a degree of certainty that a conventional risk assessment of GM crops simply cannot provide. According to them, the process merely showed that the evidence the EPA had in front of them at the time did not indicate any significant harm or hazard to the monarchs that could be associated with Bt crops. To classify Bt crops as safe for commercial growing on this basis – because there is no strong proof that they are unsafe – indicates that the regulatory agency was demonstrating what can be described as an 'evidentiary' attitude to risk and uncertainty.

Activity 4.6

Think back to the conservation case study earlier in the chapter, and in particular the precautionary attitude to risk and uncertainty introduced there. With that in mind, consider how you would characterize the difference between it and the evidentiary attitude described here, and especially its implications in terms of assigning a conservation status to a species.

Comment

Although you are unlikely to come up with this exact formulation, perhaps the simplest way to differentiate the two approaches is to consider the precautionary as looking for there to be 'evidence of absence' of a risk before something is declared 'safe', whereas the evidentiary approach requires only 'absence of evidence' of a threat before confirming safety. In terms of assigning conservation status, the precautionary approach – as we have seen – will classify a species as threatened unless it is certain that it is not threatened. An evidentiary approach, on the other hand, will classify a species as threatened only when there is strong data to support a threatened classification.

As we saw in the conservation case, it requires good quality, robust data about both probabilities and outcomes in order to arrive at a reliable calculation of risk in the technical sense. However, the NGO group that we are following here have argued that in the case of new technologies such as GM crops, precisely because they are novel, we can have little knowledge of any quantity or quality about their effects. The NGOs contend that it is impossible to predict fully the consequences associated with their introduction using just conventional risk assessment techniques. For such cases, there needs to be a recognition that we are largely dealing with uncertainties.

We have already encountered the notion of uncertainties, in the form of unknowns, earlier in this chapter. You will recall that we used the example of normal variability in connection with the monarch's over-wintering numbers to illustrate how difficult it can be to calculate the probability (and thus risk) of population extinction. It should have been clear from that example and comparable situations that it is possible to reduce such uncertainties by gathering more and better scientific information. If we had a better sense of how monarch populations fluctuate in the absence of the hazard of deforestation, then, we could move towards calculating with more confidence the risk posed by that hazard. Similarly, with GM crops we may not fully understand at present how Bt corn will affect soil micro-organisms because, to date, experiments have not been designed to detect such effects. There is no reason in principle, though, why such experiments should not be designed, carried out and thus our uncertainties about this subject at least reduced (even if they can never be eliminated). Uncertainties or unknowns of this sort, then, can effectively be treated as current but potentially temporary deficiencies in knowledge.

| hazards |
| probabilities |
| unknowns |
| attitudes |
| risk assessments |
| indeterminacy |
| trust |

However, the arguments of those criticizing the use of risk assessment as the sole way of dealing with risks and uncertainties in the Bt–monarch controversy are stronger than this and – crucially – involve an additional sort of uncertainty. Sometimes referred to as 'radical' uncertainty, we shall use the term **'indeterminacy'** for this, to distinguish it from uncertainty in the sense of unknowns. Indeterminacy is the *sixth* way in which people tend to talk about risk, uncertainty and environments; it describes the result of the inevitable gap that exists between narrow experimental conditions and the complexities of the real world.

indeterminacy

Its major implication is that we can never fully understand or predict the effects of new technologies through isolated studies because the various assumptions inherent in their design mean that their results will have only a limited relevance in the multiplicity of environments in which such technologies will actually be used. This is not meant to imply that we should never introduce any new elements into existing environments – a sort of precautionary principle taken to ridiculous extremes – but rather that when we do we need to be open to the possibility that things may turn out rather differently from what we expect.

It is precisely this openness to unintended consequences that the SEHN, Greenpeace and the UCS find lacking in the insistence on the part of the EPA that a conventional risk assessment is enough to eliminate the possibility of Bt crops acting as a threat to the monarch. According to the NGOs, by doing so the EPA have made a number of assumptions, including the following.

- It is appropriate to treat Bt corn as what is called 'substantially equivalent' to non-GM corn, the only difference being the trait imported through the genetic engineering procedure. It is on this basis that existing information about monarch behaviour in (non-GM) cornfields has been used to inform an assessment of what they will do in fields of Bt corn.

- It is appropriate to extrapolate confidently from short-term, small-scale laboratory or field experiments designed to measure acute effects in order to determine the range of possible future interactions between monarchs and Bt corn (including chronic, long-term and sub-lethal effects). It is on this basis that the small amount of data that has been generated, involving actual interactions between the butterflies and the plant, has been used to determine a response to potential harm.

- It is appropriate to accept that Bt crops are as effective in reducing pesticide use by farmers and are as economically necessary as their makers suggest. It is on the basis of these assumed benefits that any costs revealed by the risk assessment procedure have been deemed insignificant.

You can learn more about indeterminacy in **Blowers and Hinchliffe (2003)**.

The point here is not that such assumptions about the situation were and are wrong but that they need to be acknowledged when operating in conditions of indeterminacy. They reflect the particular disciplinary affiliations, institutional locations and so on of those making them, and cannot be eliminated in any meaningful sense. By not making them clear, though, it is argued that the EPA and other organizations in their position can dangerously underestimate the limits of their knowledge and thus overestimate the robustness of their decisions. Another dimension to this challenge to 'expert' knowledge and judgement is trust.

4.4 Trust

Whatever the coherence of the indeterminacy argument presented above, according to the EPA the criticisms levelled against their handling of the Bt–monarch controversy were simply wrong. And what of those many members of the public who shared the NGOs' scepticism of the claims that enough had been done to assess the environmental implications of GM crops? They were simply wrong too. A construction such as this, which protrays those outside the policy-making community as being misinformed in various ways when it comes to environmental issues, is hardly new. In fact you will probably recognize it as being a very common move by those informing and making environmental policy decisions at many levels. Whether we are talking about large-scale national events like the BSE (mad-cow) crisis or the GM controversy, or smaller-scale local ones like the building of a bypass or the location of a landfill site, the environmental concerns of the public in the face of reassuring official pronouncements are regularly dismissed out of hand.

This disregard for public opinion by those in positions of power has a long history (and not just an environmental one). So, too, has the wariness of the public when confronted with the introduction of new technologies. In Britain, people's increasing anxieties about nuclear power since the 1950s (see Chapter One) is often held up as signalling the start of a growing ambivalence to the idea of progress which such innovations were meant to embody. By all accounts the idea that not everyone was overwhelmingly in favour of this new 'fuel of the future' was met with bemusement by the people in charge of the nuclear programme at that time. Their reaction has become the standard response of innumerable groups of policy makers in similar positions ever since, and was twofold. The first aspect is to assert that the reason why the public cannot recognize the enormous benefits brought by nuclear power is simply that they do not understand the underlying science on which it is based. If they did – so the story goes – they obviously would not and could not make such 'irrational' judgements concerning its safety. The second aspect is merely a reinforcement of an existing opinion that when it comes to science and technology, even science and technology that affects everybody, 'experts know best'.

The GM controversy was only one opportunity among many for the recapitulation of this story. Since that time, however, what has changed is that social scientists have started to test and contest that position by taking an active interest in how ordinary people actually do make judgements about risky situations (including environmental ones) of various kinds. One of the main lessons from this research is that individuals and groups take decisions very much based on what they know rather than what they don't know. Early work in the area, for example, focused on the perception of risks and demonstrated that when deciding what is and what is not acceptable a number of factors seem to influence people's reactions. One example is the extent to which the risk is under the control of the individual. Another is the size of any possible catastrophe involved. According to this approach, the reasons why the public have come to fear nuclear power are that it scores badly on factors that people rate highly: its unfamiliarity, its potential to produce highly concentrated harm and the non-voluntary nature of the risk involved all count against it (Slovic, 1992).

More recent work has moved on from this rather mechanical method to become more concerned with issues of trust, and it is trust that becomes our *seventh* and final way that people tend today to talk about environmental risks and uncertainties. From this point of view, public reaction to new technologies are not just their reactions to (their understandings of) the risks involved, but also to the behaviour, track record and general trustworthiness of the institutions responsible for those new technologies. This is not to suggest that people simply make an emotional response to the question of who is responsible. Instead, such a way of assessing risk can be seen as demonstrating a very sophisticated sense on the part of the lay public of some of the uncertainties and indeterminacies which – as we have seen – almost inevitably characterize environmental issues (for more on trust and GM, see Wynne, 2002).

hazards
probabilities
unknowns
attitudes
risk assessments
indeterminacy
trust

Activity 4.7

Think back over what you have learned about trust here and your own personal experience (if any) of the GM debate in Britain. With particular reference to food issues, can you speculate on why the government, scientific and industry experts mentioned may not have seemed very trustworthy?

Comment

Although the history of 'food scares' throughout the 1980s certainly helped to build a climate of suspicion around how such matters were dealt with in Britain, it is now commonly accepted that the BSE crisis played a central role in determining how GM was received. Bombarded with a continual stream of reassurances that beef infected with this disease was safe for humans to eat, only finally to be told that people could in fact contract a human equivalent (CJD), it is perhaps hardly surprising that a new instance of government, science and industry again coming together around a controversial food issue sounded alarm bells. In fact we may well ask the question who was acting more rationally, 'the experts' or 'the public'? If you have experience of other national contexts, you might like to think about whether the British case is unusual or not.

These findings, now well substantiated, cast new light on the sorts of climates of suspicion that seem to grow periodically around particular environmentally related technologies. Public response to GM – much stronger in the British context than in the USA context we have been describing in detail here – represents a good example. Rather than regarding the furore that accompanied the arrival of GM foodstuffs in the UK food chain as the expression of an ill-informed and hysterical overreaction to a tried and tested technique – as the experts framed the situation – this response looks significantly different. Effectively, the public were not asking themselves the question: is GM food safe for human health and environments? The question now became: do we trust the government, scientific and industry bodies making the decisions with our food? The answer (certainly for the period between 1998 and 2002) was very clearly 'no'.

Protesters: WTO meeting, Seattle, December 1999.

Summary

In this section we have used the Bt corn–monarch controversy as an example of why decision makers are increasingly concerned about the environmental consequences of introducing new technologies. The case has raised three further ways in which environmental risks and uncertainties can be considered in addition to those covered in Sections 2 and 3:

- *Related to risk assessment (such as that carried out by the EPA)*
  Risk assessments are a key tool in putting into policy practice the preventative approach to environmental harm that seeks to avoid rather than clear up future problems. Although there are many different versions of the basic risk assessment procedure, they all involve the three steps of identification, estimation and evaluation.

- *Related to indeterminacy (raised by critics of the EPA's evidentiary approach)*
  Taking an evidentiary attitude to risk means being satisfied with 'absence of evidence' of a particular hazard. This can be contrasted with the 'evidence of absence' required by a precautionary attitude. Indeterminacy draws attention to the difference between experimental and real-world conditions. Hidden and often controversial assumptions can frame risk assessments.

- *Related to trust (when issues of public concern are addressed in different ways)*
  Experts often characterize people's knowledge of environmental issues in terms of a lack of proper scientific understanding. Recent work in the social sciences has shown that the public is actually quite skilful in deciding how to respond to environmental risk and uncertainties. Increasingly, it is recognized that the degree of trust or confidence that individuals and groups have in the institutions responsible for managing particular environmental risks and uncertainties plays a key role in influencing their reactions to them.

5   Conclusion: being confident about risk and uncertainty

This chapter has tried to emphasize the importance of the analytical concepts of risk and uncertainty for understanding environmental issues. We have used three approaches in order to achieve this aim. First, we have provided a sketch of the broad context in which risk and uncertainty have become very important ways of organizing environmental policy at a variety of levels. In particular, a shift to a preventative approach to environmental harm was identified in the introduction and illustrated throughout the chapter. An increasing concern with the unintended consequences of new technologies was highlighted in the second half of the chapter. Second, we have used a number of case studies connected to this book's theme of extinction and endangerment to illustrate in some detail the implications in practice of the different ways of treating risk and uncertainty. Our review of the IUCN's methodology for assigning threat statuses to endangered species and the controversy over the possible hazard posed to the monarch butterfly by certain GM crops in the USA have been the key examples here. Finally, we have pointed outwards from these themed focuses to show how the concepts of risk and uncertainty can be applied to a wide variety of environmental issues. Nuclear power, climate change and BSE have been some of those mentioned.

| hazards |
| probabilities |
| unknowns |
| attitudes |
| risk assessments |
| indeterminacy |
| trust |

You should now feel confident about two things. The first is that you can recognize the particular version (or versions) of risk and uncertainty (from our list of seven) that is operating in most environmental situations you encounter. Second, you can appreciate why disputes about what to do in a particular environmental situation are often the outcome of a difference of opinion about which version of risk and uncertainty is the most appropriate one to apply in that situation. Being able to remember three key distinctions will help you with this:

- *The distinction between hazard and risk in a technical sense*
  A hazard, you should recall, may be defined as any event, process or product that poses a threat to humans, non-humans or habitats. Hazard can tell us something about the possible outcomes of a given environmental situation, but to calculate the risk of those outcomes occurring we also require information regarding the probabilities of those outcomes.

- *The distinction between uncertainties in the technical sense and indeterminacy*
  Uncertainties in the technical sense are unknowns due to measurement problems or normal variability that are, in theory at least, more or less resolvable with more or better experiments and data. Indeterminacy on the other hand is more the result of the irreducible difficulties of knowing what relevance the assessments one makes of the consequences of a novel technique or technology will have outside the context and assumptions in which those assessments have been made.

- *The distinction between precautionary and evidentiary attitudes to risk and uncertainty*
  The former can be defined as a guide to taking action in order to avoid harm even in circumstances when that adverse outcome is not certain. The latter, by contrast, can be defined as the tendency to take remedial action in a given situation only if there is clear evidence that an adverse outcome has actually occurred.

References

Bingham, N. (2003) 'Food fights: on power, contest and GM' in Bingham, N. et al. (eds) (2003).

Bingham, N., Blowers, A.T. and Belshaw, C.D. (eds) (2003) *Contested Environments*, Chichester, John Wiley & Sons/The Open University (Book 3 in this series).

Blackmore, R. and Barratt, R.S. (2003) 'Dynamic atmosphere: changing climate and air quality' in Morris, R.M. et al. (eds) (2003).

Blowers, A.T. and Elliott, D.A. (2003) 'Power in the land: conflicts over energy and the environment' in Bingham, N. et al. (eds) (2003).

Blowers, A.T. and Hinchliffe, S.J. (2003) 'Environmental responses: radioactive wastes and uncertainty' in Blowers, A.T. and Hinchliffe, S.J. (eds) (2003).

Blowers, A.T. and Hinchliffe, S.J. (eds) (2003) *Environmental Responses*, Chichester, John Wiley & Sons/The Open University (Book 4 in this series).

Brower, L.P., Castilleja, G., Peralta, A., Lopez-Garcia, J., Bojorquez-Tapia, L., Diaz, S., Melgarejo, D. and Missrie, M. (2002) 'Qualitative changes in forest quality in a principal overwintering area of the monarch butterfly in Mexico, 1971–1999', *Conservation Biology*, vol.16, no.2, p.355.

Drake, M. and Freeland, J.R. (2003) 'Population change and environmental change' in Morris, R.M. et al. (eds) (2003).

EAA (2001) *Environmental Issue Report, No.22*, Luxembourg, OPOCE.

EPA (2001) 'Biotechnology corn approved for continued use', US Environmental Protection Agency, Press Release, 16 October.

Grace, E. (1997) *The World of the Monarch Butterfly*, San Francisco, CA, Sierra Club Books.

Hinchliffe (2000) 'Living with risk: the unnatural geography of environmental crises' in Hinchliffe, S.J. and Woodward, K. (eds) *The Natural and the Social: Uncertainty, Risk and Change*, London, Routledge.

Jasanoff, S. (1999) 'Songlines of risk', *Environmental Values*, vol.8, no.2, pp.135–52.

Lupton, B. (1999) *Risk*, London, Routledge.

Morris, R.M. (2003) 'Changing land' in Morris, R.M. et al. (eds) (2003).

Morris, R.M., Freeland, J.R., Hinchliffe, S.J. and Smith, S.G. (eds) (2003) *Changing Environments*, Chichester, John Wiley & Sons/The Open University (Book 2 in this series).

Slovic, P. (1992) 'Perception of risk: reflections on the psychometric paradigm' in Krimsky, S. and Golding, D. (eds) *Social Theories of Risk*, Westport, Praeger.

Wells, S., Pyle, R.M. and Collins, N.M. (1983) *The IUCN Invertebrate Red Book*, Switzerland, IUCN.

Willets, P. (2002) 'What is a non-governmental organization?', *UNESCO Encyclopedia of Life Support Systems*, UNESCO, www.eolss.net (accessed 31 October 2002).

WWF (2000) *Global Warming and Terrestrial Biodiversity Decline*, Switzerland, WWF.

Wynne, B. (2002) 'Creating public alienation: expert cultures of risk and ethics on GMOs', *Science as Culture*, vol.10, no.4, pp.445–82.

Conclusion

Steve Hinchliffe

As we noted at the outset, our aim in this book is to enable you to achieve a number of objectives. First, by now, we hope you will be able to answer some key questions regarding the study of environments. These include the following:

- How can we approach the study of environments?
- Why are environmental issues so pressing?
- How can we make sense of environmental issues?

Second, you should be able to demonstrate, through reference to particular places and processes, the usefulness of a number of key themes and concepts in environmental studies. This conclusion will try to draw together some of the answers to our key questions and some of the main aspects of our themes and concepts in order to realize a third objective. This is to point you forwards, enabling you to consider how the material in this book will assist you in your future engagement with environments and environmental issues.

In Chapter One you learned that a productive way of approaching the study of environments is to focus upon three themes – change, contest and response. You also learned that: environmental issues are pressing precisely because they are in a constant state of change; there are contests over understanding and actions; and there are debates over how best to respond to these changes and contests. Here we shall revisit the question of why environmental issues are so pressing (the second key question above) and draw out a number of observations that can be gleaned from the chapters in this book. We shall then go on to reflect upon the approaches that this book has taken to making sense of environmental issues (the first and third key questions above).

Why are environmental issues so pressing?

Looking back at the chapters we can find at least four concerns that seem to make environmental issues increasingly pressing.

First, there is evidence that some environments are changing in ways that may threaten their future survival. For example, some environmental processes, such as extinction, climate change, sea level rise, deforestation and coastal erosion, seem to be occurring at ever-increasing rates. In Chapters One and Two you learned that whilst processes like these have long geological histories, there is nevertheless concern that we are currently experiencing a period of rapid and possibly unsustainable change. Whilst species have been going extinct since there was life on Earth, and even though the planet's biodiversity may now be greater than at any other time in the past, the current rate at which species are

For more detail on environmental changes, the causes and consequences of change on land, ocean and in the air, see **Morris et al.** (2003).

disappearing seems much higher than we would expect from the background rate. A similar concern underlies changes in climate. Although it is true to say that global and local climates have always been changeable, many argue that the present rate of change is beyond what would be expected for the current interglacial. Certainly, in these and other cases the facts are rarely clear-cut. But the issue of change, its direction and rate, is one major reason for environmental questions being so pressing.

Our second reason for concern focuses on the consequences of environmental changes and, in particular, on their uneven distribution. Change is rarely a concern in and of itself. It is the consequences of environmental change that are significant. We know that environmental change is the norm and that, as numerous extinctions of the past testify, change often has consequences. Yet there is a perception, at least, that the consequences of contemporary changes may be particularly hard to bear. To be sure, a good part of this concern is directed at our own species. Can humans survive when they are seemingly undermining the very things that make life on the planet possible and, for some, pleasurable? This is a question that asks us to think in the longer term. Even more pressing, for many, is the view that some people – and some non-humans – are more vulnerable to change than are others, and are currently finding survival on a rapidly changing planet extremely difficult. Indeed, it is increasingly clear that the effects of environmental changes will be felt unevenly over space and in time. Some may benefit from change whilst others pay the costs, with the result that contests and conflicts are likely to become more frequent. It is worth expanding on this by way of some examples raised in the foregoing chapters.

Contest and conflict are described and explained in Bingham et al. (2003).

As was suggested in Chapter One, sea level change is neither uniform across the globe nor are its effects felt evenly. And from Chapter Three we saw that people who are already living in conditions that make it difficult to find enough food, water and other essentials are particularly vulnerable to environmental changes. Adapting to change, through, for example, managed coastal retreat or through changing agricultural practices, may be more of an option in a food-rich state such as the United Kingdom than in areas where land and food are both in short supply; thus, to some extent environmental conservation is an option available to the affluent. Meanwhile, just as there are inequities across space, so are there through time: for example, future generations may well be impoverished by actions taken today. This has been a particular feature in nuclear power and radioactive waste management debates where the cost of waste management will be felt for many generations to come whilst the benefits (in terms of electricity supplied) of power stations such as Bradwell have already been enjoyed. It is also a concern of some people that processes of extinction will deny future generations the instrumental and non-instrumental values that, say, tropical rainforests and rare species offer. Similarly, pollution of land, air and water may reduce the value of environments to future generations. So future generations may be more vulnerable than we have been accustomed to believing, something that sustainable development attempts to address (see Chapter One). Last, and by no means least, uneven vulnerability applies to non-humans too. Indeed, as the three chapters on extinction have shown, many non-human species are at high risk of being reduced to small, isolated, unsustainable populations (see especially Chapter Four).

Our third reason why environmental issues seem so pressing relates to the role that human beings play in causing environmental changes. When you learned about the acceleration in species extinctions, it was also suggested that human beings were one of the main instigators of this change in rate. Similarly, when people talk about climate change, deforestation, toxic environments or soil erosion, the finger of blame is often pointed at human beings and their societies. Whilst it is often difficult to draw a line between human and non-human causes of environmental change, it seems fairly uncontroversial to suggest that today, more than was the case in the past, human beings do affect their environments and the environments of almost all other living and non-living entities to an extent that was unimaginable only a few centuries ago. This is partly a result of the growing power of human beings to affect environments (see Chapter Three). Humans can now wipe out previously abundant non-human populations as well as alter the climate of the whole globe, and in timespans that can be shockingly short. With that degree of power comes an unprecedented responsibility. So, environmental issues are in part so pressing because people are starting to recognize the human species' role in changing the planet and the need to respond accordingly.

Environmental responses, and responsibilities, are discussed further in **Blowers and Hinchliffe (2003)**.

Environmental issues, then, are pressing because of rapid change, uneven consequences and human responsibility. To this list we can add one more aspect that became particularly evident in Chapter Four. Environments are complex, and are often therefore unpredictable. In Chapters One and Two the authors pointed out how environments are made up of interconnections: a change in one aspect often leads to changes in others. Building a sea wall in one place can have knock-on effects along the coastline, causing loss of saltmarshes or breaches elsewhere; likewise, the introduction of a new species onto an island can produce numerous population effects. In order to convey this interconnectivity, terms such as 'ecosystem' and 'food web' have been introduced. The lesson from this understanding of environments is that a particular action in one place and time can lead to a multitude of outcomes, some immediate and local, but many that have effects in other places and at other times. The result is that it is difficult to know for sure what will happen next or how a particular action will turn out. This applies to a number of areas, from breaching a sea wall, to reducing carbon dioxide emissions, to introducing genetically modified crops. Making the right decisions in conditions of such complexity is a difficult business.

The interconnectivity between living and non-living aspects of environments is described in more detail in **Morris et al. (2003)**. The effect that this has on environmental responses is explored in **Blowers and Hinchliffe (2003)**.

Rapid changes, uneven vulnerability, human responsibility and complexity: these are four elements that combine to make environmental issues so pressing. There are possibly more reasons, or ways of extending the ones that we have mentioned. In order to make sense of these and other concerns, we have made use of a number of approaches in this book. We can now turn to reflect upon these approaches.

Approaching and making sense of environmental issues

In Chapter One you were introduced to a number of themes and concepts that will help you to study environments and environmental issues. The themes were change, contest and response. These help you to structure your investigation of an environmental issue: in short, they can help you to describe what is going on in an environmental situation. Identifying how environments are changing (change), how they are matters of debate and sometimes controversy (contest), and what if anything is being done and also how the environment itself is responding (response): these are all basic elements of most environmental issues. The concepts that were introduced were space and time, values, power and action, and risk and uncertainty. They cut across many environmental issues and can be used to ask searching questions about a particular circumstance or event, or about more general environmental processes. Indeed, we have suggested that these will help you not only to describe an environmental issue but also to analyse it. To analyse is to offer more than a description of an issue. Technically it means to break something down into its component parts or to identify the principles at work. The themes of change, contest and response were introduced extensively in Chapter One and so we will have less to say about these here. We will spend the remainder of this conclusion summarizing some of the main aspects of the analytical concepts that were set out in the opening chapter and were introduced in more detail in each of the subsequent chapters.

Time and space

Our view of a coastline alters when we look back even a few hundred years and realize that the boundaries between land and sea have never been static. It also alters when we think forward more than a few years to consider how actions now will affect future generations. Time is central to so many aspects of environmental issues. Choosing the right timescale is an important consideration in environmental studies. Looking at short timescales may tell us little about the changes that are occurring in an environment. You came across this problem in Chapter Four, in the case of the monarch butterfly (*Danaus plexippus*). The population of monarchs varies substantially from year to year, so it is difficult to say, with data for only a few years, whether there is a trend in population change and whether or not that trend is significant. This is a familiar problem in environmental studies where data are sometimes relatively recent and often do not provide the required coverage. Nevertheless, an appreciation of how environments have changed in the past is vital if we are to comprehend current and future changes.

The importance of timescales is developed in Morris et al. (2003).

As Chapters One and Two also made clear, change doesn't occur at the same pace everywhere. Some places are changing faster than others. Global sea level rise has been estimated to be 0.5 metre in the coming century (though within a range from 0.09 to 0.88 metre, as you saw in Chapter One). But this does not mean that the effects will be the same everywhere. Places like the Blackwater estuary may be more severely affected because, as you learned in Chapter One, the land is already sinking as a result of regional geology. There may well be some places where global sea level rise is experienced as a drop in sea level, areas, for example, where there is local uplift. So we need to be careful how we choose our spatial as well as our temporal scales. Speaking in global terms may miss the local differences and inequalities.

But there is more to space than scale, a fact well illustrated in Chapter Two and picked up in Chapter Three. The degree to which places interconnect is also important to many environmental processes. The ability, for example, of non-human populations to survive and prosper is partly dependent upon their spatial relationships, illustrated by the detrimental effect of spatial fragmentation of habitat. If this has already occurred, then much work may go into devising wildlife corridors to allow dispersal and connectivity between isolated populations. This is one example where consideration of an issue requires giving thought not only to a particular place but also to the relationships that this place has with others. So to understand the Blackwater estuary, for example, it is necessary not only to pay attention to what is happening on the site, so to speak, but also to trace the Blackwater's links to other places. Bird migrations connect the grazing marshes to Siberia. River flows tie the saltmarshes to processes inland. Tidal movements link the estuary to what happens offshore. European Directives tie in the farms and nature reserves with Westminster and Brussels. Sea level rise links the sea defences to carbon emissions from power stations across the world. Meanwhile, connections and relationships between places can alter, and in doing so produce notable environmental effects. The railroads, for example, contributed to a rapid and disastrous environmental change for passenger pigeons: as was shown, the linking of consumers in New York to pigeon flocks in the Mid West was a major component in the bird's demise.

In sum, both space and time are important when addressing environmental issues. Consideration of past changes and future effects makes time a central aspect of environments. Space is important in part because it alerts us to the importance of considering scale, and thereby of comparing general or global trends with the uneven effects that are felt in particular places. Space is also important because it highlights the relationships between places and the effects that connections and fragmentation can have on the environments of those places.

Values, power and action

We have already established that human beings have a crucial role to play in environmental issues. They are not the only important actors, nor are they necessarily the most important actors. But it seems inescapable that what humans do and don't do affect great parts of the planet, and that only by understanding how and why people act in the ways that they do can we hope to achieve sustainable environments. The terms 'values', 'power' and 'action' have been used to allow us to explore what drives and directs human endeavours. We have learned that people can value things as means to other ends, a so-called instrumental form of value. But environments and aspects of environments can also be valued as ends in themselves. By broadly considering what people value, we might get closer to understanding people's actions. Understanding actions and their relation to what people do or don't consider important is a key step if we are to start to address environmental issues.

We have also learned that actions are not just about desires and values. What we do is also coloured, and to a greater or lesser extent shaped, by power relations. It is not always possible to choose to do what you think is the right thing, simply because you are not allowed to choose or because not enough resources can be found for your action to take effect. Meanwhile, it is clear that power is unevenly distributed: some people, groups and societies have more power at their disposal than others; they can mobilize resources and may be able to shape the discourses that make their actions more feasible. Power as a focus of study can therefore help us to explain why some actions are more likely to succeed than others. Finally, by understanding the distribution of power and considering how such a distribution may be challenged or altered through actions, we can start to think about changing environmental futures.

Risk and uncertainty

The interconnected and complex nature of environments and the level of human responsibility for environments combine to make risk and uncertainty a feature of many environmental issues. It is often difficult – if not impossible – to know exactly what effects any particular action will have in the future. From introducing genetically modified corn to reducing CO_2 emissions, the effects can at best only be estimated. In short, environmental issues are characteristically issues where uncertainties, or unknowns, figure prominently. This is one reason why decision-making guidelines such as the precautionary principle, introduced in Chapter One and explained further in Chapter Four, have become so important to environmental policy-makers.

We use risk, in its more technical sense, to measure the probabilities of an action leading to a particular outcome. Building a sea wall, for example, may entail a high risk of saltmarsh depletion. Constructing a dam may reduce the risk of failures of irrigation waters but increase the risk of tiger extinction in rural India. In a less technical, more sociological sense, risk has become a feature of many parts of contemporary life. Risk is essentially about asking questions of the future: what will happen if ...? A number of sociologists have recently characterized modern society as a 'risk society', largely because such future-orientated questions have become more and more common. In the past, they argue, people tended to get on with things whereas, today, everything becomes a project where the future has to be taken care of rather than being allowed to take care of itself. As ideas such as sustainable development testify, future-orientated questions are characteristic of environmental issues. So, in its technical and more general sense, risk has become a characteristic feature of environmental issues.

If uncertainty and risk are characteristic of environmental issues then their analytical importance becomes important precisely because they are so often calculated, evaluated and used in different and sometimes conflicting ways (as Chapter Four indicated). Whilst the risk of being killed crossing a road may be higher than that of contracting a radiation-related illness from nuclear waste disposal, many would argue that these measures are not comparing like with like. As you learned in Chapter Four, risk is as much about choice and trust as it is about probabilities. People's ability to choose whether or not they face particular risks affects their willingness to accept a risk. Acceptability of risk is also about the degree to which people trust the institutions that are placed in charge of a particular hazardous activity. Likewise uncertainties are often handled in different ways – they are underestimated or overestimated to suit particular purposes. Sometimes they are suggested to be temporary, requiring nothing more than new research, whilst at other times they are regarded as permanent unknowns. In short, the ways in which risk and uncertainty are discussed reveals much about the aims and values of those discussing them. It becomes important, then, in any environmental situation to analyse how risks and uncertainties are being estimated, described, highlighted or ignored.

✳   ✳   ✳

Our aim has been to equip you with a series of starting points for approaching environments and environmental issues. The themes and concepts that you have seen developed in this book and have practised using in the activities and questions will, we hope, enable you to make that approach exciting, creative and informative. In later books in this series you will be able to explore many issues, from wind farms to fish farms, from water wars to nuclear waste dumps, and from soil science to environmental mass media. There, and we hope elsewhere in your studies and in your day-to-day engagement with environmental issues, you will be able to practise and hone this analytical approach to become a skilled, and multi-disciplinary, environmental practitioner.

References

Bingham, N., Blowers, A.T. and Belshaw, C.D. (eds) (2003) *Contested Environments*, Chichester, John Wiley & Sons/The Open University (Book 3 in this series).

Blowers, A.T. and Hinchliffe, S.J. (eds) (2003) *Environmental Responses*, Chichester, John Wiley & Sons/The Open University (Book 4 in this series).

Morris, R.M., Freeland, J.R., Hinchliffe, S.J. and Smith, S.G. (eds) (2003) *Changing Environments*, Chichester, John Wiley & Sons/The Open University (Book 2 in this series).

Acknowledgements

Grateful acknowledgement is made to the following sources for permission to reproduce material within this book.

Figures

Chapter One contents page: © Bob Glover; *page 4:* © Bob Glover; *Figure 1.1:* © Darren Wycherley; *Figures 1.2, 1.8 and 1.29:* Crown © reproduced with the permission of Ordnance Survey on behalf of Her Majesty's Stationery Office, Licence No ED 100020607; *Figure 1.4:* © Darren Wycherley; *Figure 1.5:* © The Environment Agency, Eastern Area; *Figure 1.6:* © The Environment Agency; *Figure 1.7:* © The Environment Agency; *Figure 1.9:* © Reproduced by courtesy of the Essex Record Office; *Figure 1.10:* Ziegler, P. (1997) The Black Death, Curtis Brown on behalf of Philip Ziegler, Copyright © Philip Ziegler 1997; *Figure 1.11:* © The Art Archive/Biblioteca Nazionale Marciana, Venice/Dagli Orti; *Figure 1.12:* Lucy, G. (1999) Essex Rock: A Look Beneath the Essex Landscape, Essex Rock & Mineral Society; *Figure 1.13:* © Darren Wycherley; *Figure 1.14:* adapted from Houghton, J. T., Jenkins, G. J. and Ephraume, J. J. (eds, 1990) Climate Change The IPCC Scientific Assessment, Cambridge University Press, © Intergovernmental Panel on Climate Change; *Figures 1.15 and 1.16:* Hunter, J. (1999) The Essex Landscape: A Study of its Form and Histv ory, Essex Record Office; *Figure 1.17:* Lucy, G. (1999) Essex Rock: A Look Beneath the Essex Landscape, Essex Rock & Mineral Society; *Figure 1.18:* Based on Murray, J. W. (1992) Palaeogene and Neogene In: Atlas of Palaeogeography and Lithofacies (eds. J. C. W. Cope, J. K. Ingham and P. F. Rawson) Memoir 13, pp. 141-147, Geological Society, London; *Figure 1.19:* © Mark Edwards/Still Pictures; *Figure 1.20:* © Darren Wycherley; *Figure 1.21:* © Crown copyright, 2002. Reproduced by permission of CEFAS, Lowestoft; *Figure 1.22:* © Courtesy of BNFL Magnox Electric plc; *Figure 1.23:* © Times Newspapers/News International; *Figure 1.24a:* R.S.P.B. East Anglian Regional Office; *Figure 1.24b:* © Bob Glover; *Figure 1.25b:* © Bob Glover; *Figure 1.26:* R.S.P.B.; *Figure 1.27:* East Anglian Salt Marshes, Environment Agency; *Figure 1.28:* © Popperfoto; *Figure 1.30:* © The Environment Agency, Eastern Area; *Figure 1.31:* © The Environment Agency, Eastern Area; *Figure 1.32:* Essex Wildlife Trust; *Figure 1.33:* © The Environment Agency, Eastern Area.

Chapter Two contents page: © Kelvin Conrad; *Figure 2.1a:* © Richard Brooks/ FLPA; *Figure 2.1b:* © W.Wisniewski/FLPA; *Figure 2.1c:* © Hans Gebuis; *Figure 2.1d:* © Hans Gebuis; *Figure 2.2a:* © Kenneth W.Fink/Ardea Diorama in the Colorado Museum of Natural History, Denver, USA; *Figure 2.2b:* © Courtesy of The Native Fish Conservancy of North America; *Figure 2.2c:* © Dave Watts/ NHPA; *Figure 2.2d:* © Courtesy of the Royal Botanic Gardens Kew, and Bonn Botanic Gardens; *Figure 2.4:* © Frances Fawcett; *Figure 2.5a:* © Tim Davies/ Science Photo Library; *Figure 2.5b:* © Judd Cooney/Oxford Scientfic Films;

Figure 2.5c: Courtesy of the United States Fish and Wildlife Service, Division of Endangered Species; *Figure 2.5d:* Courtesy of Steve Compton, Leeds University © Photo: Roger Key/ English Nature; *Figure 2.7:* © Peter Sheldon. From: S193 Fossils and the History of Life, Figure 4.4c, page 84; *Figure 2.8:* © Pascal Goetgheluck/Science Photo Library; *Figure 2.10:* © John Watson/OU; *Figure 2.11a:* Heather Angel; *Figure 2.11b:* John Mason/Ardea; *Figure 2.13a:* © Tom Willock/ARDEA; *Figure 2.13b:* © Mike Dodd; *Figures 2.16, 2.17 and 2.19:* Groombridge, B. and Jenkins, M.D. (2000) Global Biodiversity: Earth's Living Resources in the 21st Century, World Conservation Monitoring Centre; *Figure 2.18a:* © Martin Wendler/NHPA; *Figure 2.18b:* © Martin Wendler/NHPA; *Figure 2.20a:* ©Thierry Montford/Still Pictures; *Figure 2.20b:* © Roland Seitre/ Still Pictures; *Figure 2.20c:* © Tui De Roy/Oxford Scientific Films; *Figure 2.20d:* © M & P Fogden/Fogden Photos; *Figure 2.21:* © Mark Edwards/Still Pictures; *Figure 2.24:* © D.Parer and E.Parer-Cook/Ardea; *Figure 2.25a:* © Bill Coster/ NHPA; *Figure 2.25b:* ©David Middleton/NHPA.

Chapter Three contents page: From 'Land of the Tiger' by Valmik Thapar, 1997, London, BBC Books. © Valmik Thapar; *Figure 3.1:* © Smithsonian Institution, National Museum of Natural History, Division of Birds; *Figure 3.4:* Painting by F. Bartoli © Peter Newark's American Pictures; *Figure 3.5:* © Peter Newark's American Pictures; *Figure 3.6:* From Nature's Metropolis: Chicago and the Great West by William Cronon. Copyright © 1991 by William Cronon. Used by permission of W.W.Norton & Company, Inc.; *Figures 3.7 and 3.8:* Earthtoons: The First Book of Eco-humor by Stan Eales, Grub Street Publishing; *Figure 3.9:* © Manchester Evening News Syndication; *Figure 3.10:* © Bob Bennett/OSF; *Figure 3.11:* Seidensticker, J., Christie, S. and Jackson, P. (Eds.) Riding the Tiger, 1999 published by Cambridge University Press; *Figure 3.13:* © David Woodfall/ NHPA; *Figure 3.14:* ©ANT/NHPA; *Figure 3.15:* Figure 3.14: Map courtesy of the National Geographic Society; *Figure 3.16:* Wikramanayke et al (1998) Conservation Biology, Vol 12, No 4 Page 869, Blackwell Publishers Ltd; *Figure 3.17:* (c) Zig Leszcynski/AA/ OSF; *Figure 3.18:* Sanctuary Magazine www.sanctuaryasia.com; *Figure 3.19:* © Glyn Williams;

Chapter Four contents page: © James L.Amos/Still Pictures; *Figures 4.1 and 4.3:* © Frans Lanting/Minden Pictures/FLPA; *Figures 4.2 and 4.4:* Journal of the Lepidopterists Society, Vol. 49, No 4, 1995; *Figure. 4.5:* Brower, L.P. et al Quantitative changes in forest quality in a principal overwintering area of the monarch butterfly in Mexico 1971-1999 (2002) Conservation Biology, Vol. 16 No. 2, Blackwell Science; © Lincoln P Brower; *Figure 4.7:* © Kevin T Karlson; *Figure 4.8:* Courtesy of the Department of Crop Sciences, University of Illinois, Urbana, Illinois; *Figure 4.9:* This figure was used from the Monarchs in the Classroom MonarchLab website (www.monarchlab.umn.edu http://www.monarchlab.umn. edu/) and is used with permission; *Figure 4.11:* Union of Concerned Scientists; *Figure 4.12:* Keep Nature Natural; *page 160:* © Nick Cobbing.

Tables

Tables 4.1 and 4.2: International Union for Conservation of Nature and Natural Resources; *Table 4.4*: Late lessons from early warnings: the precautionary principle from European Enviornment Agency © EEA, Copenhagen 2001.

Cover illustrations

From left to right: From 'Land of the Tiger' by Valmik Thapar, 1997, London, BBC Books. © Valmik Thapar; © Bob Glover; © N. Dickinson/Still Pictures; © James L. Amos/Still Pictures.

Index